UPKAR'S

OBJECTIVE CIVIL ENGINEERING

[For competitive selection examinations of Railway Service, Engineering (diploma) level, State Public Service Commissions, C.P.W.D., P.W.D., Coal India, Irrigation Departments and other equivalent services.]

By

Pramod Kumar Mishra

UPKAR PRAKASHAN, AGRA–2

Publishers

UPKAR PRAKASHAN

(An ISO 9001 : 2000 Company)

2/11A, Swadeshi Bima Nagar, AGRA–282 002
Phone : 4053333, 2530966, 2531101
Fax : (0562) 4053330, 4031570
E-mail : care@upkar.in, **Website :** www.upkar.in

Branch Offices :

4845, Ansari Road, Daryaganj,
New Delhi—110 002
Phone : 011–23251844/66

1-8-1/B, R.R. Complex (Near Sundaraiah Park, Adjacent to Manasa Enclave Gate),
Bagh Lingampally,
Hyderabad—500 044 (A.P.)
Phone : 040–66753330

Pirmohani Chowk,
Kadamkuan,
Patna—800 003
Phone : 0612–2673340

28, Chowdhury Lane,
Shyam Bazar,
Near Metro Station,
Kolkata–700004 (W.B.)
Phone : 033–25551510

ISBN 13 : 978-81-7482-798-2

Price : ₹ 115·00
(Rs. One Hundred Fifteen Only)

Code No. 965

Printed at : Repro Knowledgecast Limited, Thane

PREFACE

Success in examinations depends on proper planning of studies and appropriate selection of study materials. The pattern of examinations become tough. This is the reason that a right choice of study materials plays a very important role. This book cover thoroughly all the basics of the whole course as well as present to the examinee a wide spectrum of the multiple choice questions having a huge variety. The author has made a sincere attempt in this direction in the present book. Various unique features of the book are as under for example :

- A brief review of concepts at a glance covering all fundamentals and important conclusions are given at the start of every chapter.
- Chapters are classified under different units.
- Multiple choice questions in every chapter are arranged in a systematic and sequencial way covering the whole text and spectrum of the chapter.
- Answers are provided at the end of every chapters.
- **Model Test Papers** covering the whole syllabus are also provided at the end of the book again with their answers. These papers will prove to be fit for examination and provide a chance to students in assessing their level of preparation.

The present book is self-sufficient in all respects.

I am thankful to my wife Mrs. Rita Mishra who has put a hard labour in reading the proofs thoroughly and pointing out errors and omissions. My sincere thanks are also to publisher Mr. Mahendra Jain who gave me a chance to write this type of books. This edition is a nice form.

Although all attempts have been made to avoid errors and printing mistakes, but omissions are a human weakness and, therefore, constructive suggestions, modifications and errors brought to my notice will be highly appreciated and incorporated in the next edition.

—Pramod Kumar Mishra

CONTENTS

Civil Engineering

Chapter-1

APPLIED MECHANICS

Physical Quantity

All quantities in terms of which laws of physics can be expressed and which can be measured directly or indirectly are called physical quantities. Such as length, mass, time, temperature, force, velocity etc.

Unit

The chosen standard of same kind taken as reference to measure a physical quantity is called the unit of that quantity.

Unit

Fundamental units

Fundamental units are those units, which can neither be derived from one another nor can they be further resolved into any other units, *i.e.*, for mass, length and time.

Derived units

The units of all such physical quantities which can be expressed in terms of the fundamental units of mass, length and time are called Derived units such as area, acceleration, force, etc.

Mass of body

Mass of a body is the quantity of matter it contains. It is an essential property of a material body. Unit of mass is **kilogram**.

Length of an object

Length of an object may be defined as the distance of separation between its two ends. Units of length is **metre**.

Time

Time is simply what a clock reads. The idea of passage of time occurred to man from the motion of the moon across the sky and the rotation of earth about its own axis and around the Sun. Unit of time is **second**.

Four system of units are :

(1) C.G.S. system

(2) F.P.S. system

(3) M.K.S. system

(4) S.I. system

Dimension

The dimension of a physical quantity are the powers to which the fundamental units of mass, length and time have to be raised to obtain its units.

Dimensional equation

The relation bet-ween the physical quantity and its dimension is called dimensional equation.

$$F = [MLT^{-2}]$$

Dimensional constant

The quantities which possess dimensions and have a constant values are dimensional constants such as gravitational constant, Planck's constant etc.

Non-dimensional constant

The constant quantities having no dimensions are called non-dimensional constants. This includes pure numbers, trigonometrical functions, exponential functions etc.

S. No.	Physical quantity	Formula	Dimensions	S.I. units
1.	Velocity	$\frac{\text{Displacement}}{\text{Time}}$	$[LT^{-1}]$	m/s
2.	Acceleration	$\frac{\text{Velocity}}{\text{Time}}$	$\frac{[LT^{-1}]}{[T]} = [LT^{-2}]$	m/s^2
3.	Force	Mass × Acceleration	$[MLT^{-2}]$	kg m/s^2-(N)
4.	Work	$F\,r \cos\theta$	$[ML^2T^{-2}]$	kg m^2/s^2 (joule)
5.	K.E.	$\frac{1}{2}mv^2$	$[ML^2T^{-2}]$	kg m^2/s^2 (joule)
6.	Potential energy	mgh	$[ML^2T^{-2}]$	kg m^2/s^2 (joule)
7.	Torque	$Fr \sin\theta$	$[ML^2T^{-2}]$	kg m^2/s^2 (joule)
8.	Power	$\frac{\text{Work}}{\text{Time}}$	$[ML^2T^{-3}]$	kg m^2/s^3 = J/s = watt (W)
9.	Momentum	Mass × Velocity	$[MLT^{-1}]$	kg m/s or N-s
10.	Impulse	Force × Time	$[MLT^{-1}]$	kg m/s or N-s
11.	Angle	$\frac{\text{Arc}}{\text{Radius}}\left(\frac{L}{L}\right)$	$[M^0L^0T^0]$	No units
12.	Strain	$\frac{\Delta L}{L} = \left[\frac{L}{L}\right]$	$[M^0L^0T^0]$	No units
13.	Stress	Restoring force/Area	$[ML^{-1}T^{-2}]$	N/m^2
14.	Modulus of elasticity	$\frac{\text{Stress}}{\text{Strain}}$	$[ML^{-1}T^{-2}]$	N/m^2
15.	Pressure	$\frac{\text{Force}}{\text{Area}}$	$[ML^{-1}T^{-2}]$	N/m^2
16.	Frequency	$\frac{1}{\text{Time period}}$	$\frac{1}{[T]} = [M^0L^0T^{-1}]$	Hz
17.	Angular velocity	$\frac{\text{Angle}}{\text{Time}}$	$\frac{1}{[T]} = [M^0L^0T^0]$	rad/s
18.	Moment of inertia	$I\omega$	$[ML^2T^{-1}]$	kg m^2/s or J-s
19.	Surface energy	$\frac{\text{Energy}}{\text{Area}}$	$[ML^0T^{-2}]$	N/m
20.	Surface tension	$\frac{\text{Force}}{\text{Tension}}$	$[ML^0T^{-2}]$	N/m
21.	Spring constant	$\frac{F}{x}$	$[MT^{-2}]$	N/m

Force

Force is a push or pull which try to change or changes the state of rest or uniform motion of a body.

Action of Force on a Body

1. **Body at rest**—(a) If body remains at rest, force is trying to change the state of rest :

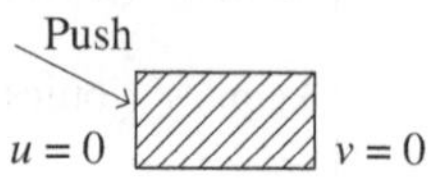

(b) If body moves, force is changing the state of rest.

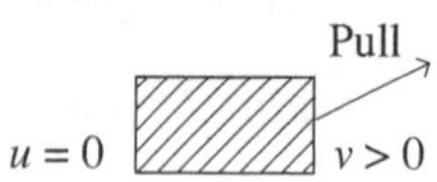

2. **Body in motion**—(a) When force is parallel or anti parallel to motion, it changes the magnitude and not the direction of motion (in a small interval of time).

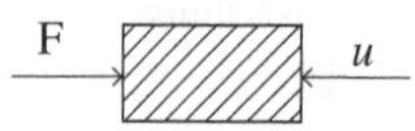

Magnitude of u increases

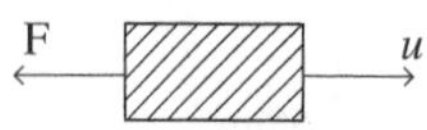

Magnitude of u decreases

(b) When force is always perpendicular to motion. Direction of motion only changes and not the magnitude and motion is uniform circular.

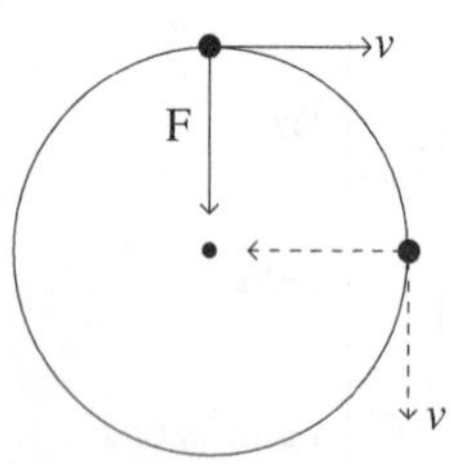

(c) Both magnitude and direction of motion changes and motion is non-uniform circular, elliptic, parabolic or hyperbolic.

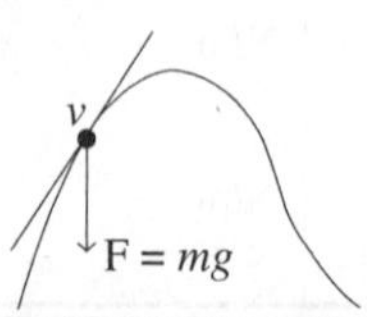

When force changes the state of rest or motion of a body, by Newton's second law

$$\vec{F} = \frac{d}{dt}\vec{P}$$

$$\vec{F} = \frac{d}{dt}(m\vec{v}) \quad (\because \vec{P} = m\vec{v})$$

$$\vec{F} = m\frac{d\vec{v}}{dt} + \vec{v}\frac{dm}{dt}$$

Since m = constant, then $\frac{dm}{dt} = 0$

$$\vec{F} = m\frac{d\vec{v}}{dt}$$

$$\vec{F} = m\vec{a} \quad \left(\because \vec{a} = \frac{d\vec{v}}{dt}\right)$$

Force = Mass × Acceleration

Law of Forces

(a) Law of parallelogram of forces

If two forces acting simultaneously on a particle be represented by the two adjacent sides of a parallelogram in magnitude and direction, their resultant may be represented in magnitude and direction by the diagonal of the parallelogram. Which passes through their point of intersection.

(b) Law of triangle of forces

If two forces acting simultaneously on a particle be represented in magnitude and direction by the two sides of a triangle, taken in order, their resultant will be represented by the third side of the triangle taken in opposite order.

(c) Law of polygon of forces

If a number of forces acting simultaneously on a particle be represented in magnitude and direction by the sides of a polygon, taken in order, the resultant of all these forces may be represented in magnitude and direction by the closing side of the polygon, taken in opposite order.

(d) Equilibrium of forces

If the net effect of various forces acting on a body is zero and they do not reduce to a couple, then those forces are to be in equilibrium.

Moment of inertia

The moment of inertia of a body about an axis is equal to the product of mass of the body and square of the distance from that axis. It is represented by I.

Table : M.I. of Most Common Sections

Section	I_{XX}	I_{YY}	I_{ZZ}
1. Rectangular section	$\frac{1}{12}bd^3$	$\frac{1}{12}db^3$	
2. Circular section	$\frac{\pi d^4}{64}$	$\frac{\pi d^4}{64}$	$\frac{\pi d^4}{32}$
3. Hollow rectangular section	$\frac{1}{2}(BD^3 - db^3)$	$\frac{1}{2}(DB^3 - db^3)$	—
4. Triangular section	$\frac{bh^3}{12}$		
5. Trapezoid	$\frac{h^3}{36}\left(\frac{a^2 + 4ab + b^2}{a + b}\right)$		
6. Square	$\frac{d^4}{12}$	$\frac{b^4}{12}$	

Friction

The opposite force which comes into play in between the surfaces of two bodies when one body moves, slids or rolls over the other body is known as force of friction or friction,

The friction observed by a body at rest is known as static friction. The friction observed by a body, when in motion is known as dynamic or kinetic friction.

Limiting friction

The maximum static friction observed by a body is known as limiting friction.

Laws of Friction

(i) The force of friction always acts in a direction, opposite to that in which the body is moving or tends to move.

(ii) The magnitude of the force of friction is equal to the force which just makes the body to move.

(iii) The magnitude of the limiting friction bears a constant ratio to the normal reaction between the two surfaces.

(iv) The force of friction is independent of area and shape of the surface in contact so long as the intensity of the normal reaction remains constant.

(v) The force of friction depends upon the roughness of the surface in contact and the material of which the surface are made.

(vi) For moderate speeds, the force of friction remains constant but it decreases slightly with the increase of speed.

Angle of friction

When a body is at the point of limiting equilibrium, the force of friction is maximum. The angle which the resultant of the maximum force of friction and the normal reaction makes with the normal reaction, is called the angle of friction and is denoted by λ.

$$\tan \lambda = \frac{\text{Maximum force of friction}}{\text{Normal reaction}}$$

Angle of repose

The angle which an inclined rough surfaces makes with the horizontal when a body placed on it, is on the point of moving down, is called angle of repose.

Periodic motion

When a body repeats its motion again and again after a fixed period, its motion is called periodic motion such as motion of earth around the sun.

Simple harmonic motion

When a particle moves right to left or up and down from a fixed point on a straight line in such a way that at any instant its acceleration is proportional to its displacement from the fixed point and is always directed towards the fixed point, then the motion of particle is said to be simple harmonic motion.

Simple pendulum

When a heavy bob suspended by a weightless inextensible but flexible string from a rigid support so that it may oscillate without any friction is called a simple pendulum.

Laws of Simple Pendulum

(i) **Law of isochronism**—For a fixed length of the pendulum and at a certain place each oscillation takes equal time if the amplitude is small (less than 4°).

(ii) **Law of length**—At a certain place the time period of pendulum is proportional to the square root of its effective length.

(iii) **Law of acceleration due to gravity**—For a certain length of the pundulum its time period is inversely proportional to the square root of acceleration due to gravity at that place.

(iv) **Law of mass**—The time period of a pendulum of fixed length and at a certain place does not depend upon the mass or material of the bob.

Compound pendulum

Compound pendulum is that rigid body which oscillates independently about an axis passing through itself. It is a practical pendulum.

Kater's pendulum

Kater's pendulum is a special type of compound pendulum by which acceleration due to gravity is determined correctly. It is a bar of copper and steel.

Gravity

The force of attraction on objects due to earth is called gravity.

Gravitation

The attraction between two bodies due to their masses is called gravitation.

Newton's law of gravitation

Newton's law of gravitation states that everybody in this universe attracts every other body with a force which is directly proportional to the product of their masses and inversely proportional to the square of the distance between their centres.

Inertial mass

Inertial mass is the mass of the material body, which measures its inertia and is defined as the ratio of magnitude of external force applied on it to magnitude of acceleration produced in it.

Gravitational mass

Gravitational mass of the material body which determines the gravitational pull acting upon it.

Kepler's law

Kepler discovered the following three laws concerning the motion of planets :

(i) **First law**—Each planet revolves around the sun in an elliptical orbit with the sun as one focus of the ellipse.

(ii) **Second law**—The straight line joining the sun and the planet sweeps out equal area in equal time intervals.

(iii) **Third law**—The square of the periodic times of the planet is proportional to the cube of the semi major axis of its orbit, *i.e.*, $T^2 \propto r^3$

Law of Conservation of Angular Momentum

It states that if no external torque acts on a rigid body, then the product of its moment of inertia and the angular velocity about the axis of rotation must remain constant.

Applications—(i) The angular velocity of a planet revolving in an elliptical orbit around the sun increases when it comes nearer the sun and vice-versa.

(ii) An ice-skater or a ballet-dancer can increase her angular velocity by folding her arms and bringing the stretched leg close to the other leg.

(iii) The speed of the inner layers of the whirlwind about its axis in a tornado is alarmingly high.

(iv) The speed of rotation of a person with some weights in his hands and a rotating turn table increases as he drows the weights to his chest.

Elasticity

The property of matter by virtue of which a body tends to regain its original shape and size after the removal of deforming forces is called elasticity.

Stress

The restoring force per unit area set up inside the body is called stress and is measured by the magnitude of the deforming force acting on unit area of the body when the equilibrium is established.

Strain

The relative change in dimensions or shape of a body which is subjected to stress is called strain and is measured by the ratio of change in length to the original length.

Hook's law

It states that under elastic limit the stress is proportional to the strain.

$$\text{Stress} \propto \text{Strain}$$

$$\text{Stress} = \text{E Strain}$$

$$\Rightarrow \quad E = \frac{\text{Stress}}{\text{Strain}}$$

where E = Constant

Work

Work is said to be done by a force if the direction of displacement is same as the direction of applied force.

Measurement of work

Work done in moving a body is measured by the product of force exerted on body and the distance moved by the body in the direction of force.

Energy

If a body has the ability to do work the body is said to possess energy.

Energy of a body is defined as its ability to do work.

Kinetic energy

The energy possessed by a body by virtue of its motion is called kinetic energy.

Potential energy

The energy possessed by a body by virtue of its position and configuration is called potential energy.

Power

Power is defined as the rate of doing work.

Law of conservation of energy

Law of conservation of energy states that energy can neither be created nor be destroyed. It can only be transformed from one form to an equivalent form.

Internal energy

An object possesses internal energy because of its temperature. The molecules of any object possess potential energy due to their locations against the intermolecular forces. Internal energy is the sum of these energies.

Heat energy

An object possesses heat energy due to the disorderly motion of its molecules.

Chemical energy

An object possesses chemical energy due to chemical binding of its atoms. Such an object may be preferably called as a chemical compound, which has lesser energy than that possessed by the parts of which it is made. This difference in energy is called chemical energy.

Electrical energy

Work has to be done in order to move an electric charge from one point to another in electric field or for the transverse motion of a current carrying conductor inside a magnetic field. This work done appears as the electrical energy of the system.

EXERCISE

1. The S.I. unit of pressure is—
   (A) Atmosphere (B) Bar
   (C) Pascal (D) mm of Hg

2. The S.I. unit of energy is—
   (A) Erg (B) Calorie
   (C) Joule (D) Electron volt

3. The dimensional formula for surface tension is—
   (A) $[MT^2]$ (B) $[MT^{-2}]$
   (C) $[M^{-1}L^0T^{-2}]$ (D) $[ML^2T^{-2}]$

4. The expression $[ML^2T^{-2}]$ represents—
   (A) Power (B) Kinetic energy
   (C) Momentum (D) Pressure

5. The dimensions of torque are—
   (A) $[ML^2T^{-2}]$ (B) $[MLT^{-2}]$
   (C) $[ML^{-1}T^{-2}]$ (D) $[ML^{-2}T^{-2}]$

6. The dimensions of pressure are—
   (A) $[MLT^{-2}]$ (B) $[ML^{-1}T^{-2}]$
   (C) $[ML^{-1}T^{-2}]$ (D) $[MLT^2]$

7. Which of the following is a scalar quantity ?
   (A) Energy (B) Momentum
   (C) Torque (D) Impulse

8. One kilogram weight in absolute units is equal to—
   (A) 32·2 dyne (B) 9·81 newton
   (C) 32·2 newton (D) 981 dyne

9. Force is—
(A) Scalar quantity
(B) Linear quantity
(C) Vector quantity
(D) Non-measurable quantity

10. The component of a force (P) at right angles to its direction will be—
(A) Zero (B) Half
(C) Same (D) 1·414 P

11. If P and Q are the force acting at an angle θ and their resultant make an angle α with P then—
(A) $\tan \alpha = \dfrac{Q \sin \theta}{P + Q \cos \theta}$
(B) $\tan \alpha = \dfrac{Q \sin \theta}{P + Q \sin \theta}$
(C) $\tan \alpha = \dfrac{Q \sin \theta}{P + Q \tan \theta}$
(D) $\tan \alpha = \dfrac{Q \sin \theta}{Q + P \cos \theta}$

12. Newton's second law of motion gives the measure of—
(A) Acceleration
(B) Force
(C) Momentum
(D) Angular momentum

13. Impulse has the same unit as—
(A) Work (B) Force
(C) Momentum (D) Power

14. When a spring is loaded, the strain produced is—
(A) Longitudinal (B) Volumetric
(C) Shearing (D) None of these

15. The change in the shape of a regular body is due to—
(A) Bulk strain
(B) Shearing strain
(C) Longitudinal strain
(D) None of the above

16. Which one of the following substances has the largest value of Young's modulus ?
(A) Rubber (B) Nylon
(C) Iron (D) Copper

17. If two forces of magnitude Pact at an angle θ, then resultant will be—
(A) $2P \cos \dfrac{\theta}{2}$ (B) $2P \sin \theta$
(C) $2P \cos \theta$ (D) $P \cos 2\theta$

18. If the resultant of two forces has the same magnitude as either of the forces, then the angle between the two forces is—
(A) 30° (B) 60°
(C) 90° (D) 120°

19. The angle between two forces, when the resultant is maximum and minimum respectively are—
(A) 0 and 180 (B) 180 and 0
(C) 90 and 180 (D) 90 and 0

20. The set of forces, whose resultant is zero, are known as—
(A) Equilibrium forces
(B) Collinear forces
(C) Coplanar forces
(D) Concurrent forces

21. The forces whose line of action lie on the same line are known as—
(A) Concurrent forces
(B) Coplanar forces
(C) Equilibrium forces
(D) Collinear forces

22. The necessary condition of equilibrium of a body is—
(A) $\sum H = 0$ (B) $\sum V = 0$
(C) $\sum M = 0$ (D) All of the above

23. Concurrent forces are those forces whose lines of action—
(A) Lie on the same line
(B) Meet at one point
(C) Meet on the same plane
(D) None of the above

24. In S.I. units, the unit of moment is—
(A) kg-m
(B) Newton-metre (N-m)
(C) N-m/sec.
(D) kg. wt.

25. The point through which the whole body weight acts, irrespective of the position of the body is called—
(A) Centroid
(B) Centre of gravity
(C) Both (A) and (B)
(D) None of the above

26. The plane figures like triangle, circle etc. have only areas, but no mass. The centre of such figures is called—
(A) Centroid
(B) Centre of gravity
(C) Both (A) and (B)
(D) None of these

27. The centre of gravity of a hemisphere of radius r from its base measured along vertical radius is—
(A) $\frac{4r}{3}$ (B) $\frac{3r}{2}$
(C) $\frac{3r}{8}$ (D) $\frac{4r}{7}$

28. The moment of inertia of a triangle of base b and altitude h with respect to a centroidal axis parallel to its base would be—
(A) $\frac{bh^3}{12}$ (B) $\frac{bh^3}{18}$
(C) $\frac{bh^3}{36}$ (D) $\frac{bh^3}{3}$

29. The moment of inertia about the base of a triangle of mass m and height h is—
(A) $\frac{mh^2}{4}$ (B) $\frac{mh^2}{36}$
(C) $\frac{mh^2}{12}$ (D) $\frac{mh^2}{6}$

30. The moment of inertia of a sphere of radius R and mass M about an axis tangential to it is—
(A) $\frac{2}{3}MR^2$ (B) $\frac{2}{5}MR^2$
(C) $\frac{7}{5}MR^2$ (D) $\frac{7}{3}MR^2$

31. I_{XX} or I_{YY} of a hollow circular section of internal diameter 'd' and external diameter 'D' will be—
(A) $\pi/32\ (D - d)^4$ (B) $\pi/32\ (D^4 - d^4)$
(C) $\pi/64\ (D - d)^4$ (D) $\pi/64\ (D^4 - d^4)$

32. If F is the limiting friction, R = normal friction then coefficient of friction μ will be equal to—
(A) $\mu = F + R$ (B) $\mu = F - R$
(C) $\mu = F/R$ (D) $\mu = R/F$

33. Coefficient of friction depends upon—
(A) Nature of surface only
(B) Area of contact only
(C) Both A and B
(D) None of these

34. M.I. of a square of side b about an axis through its centre of gravity is—
(A) $\frac{b^3}{3}$ (B) $\frac{b^4}{3}$
(C) $\frac{b^4}{12}$ (D) $\frac{b^4}{8}$

35. Static friction is always—
(A) Greater than dynamic friction
(B) Less than dynamic friction
(C) Equal to dynamic friction
(D) None of the above

36. A ladder is resting on a smooth ground and leaning against a rough vertical wall, the force of friction acts—
(A) Towards the wall at its upper end
(B) Downwards at its upper end
(C) Upward at its upper end
(D) Away from the wall at its upper end

37. In actual machines mechanical advantage is—
(A) Unity
(B) Less than unity
(C) Less than velocity ratio
(D) Greater than velocity ratio

38. The main advantage of using worm gears is—
(A) Ease of manufacture
(B) High velocity ratio
(C) Low power loss
(D) Minimum cost

39. The efficiency of a screw jack is equal to—
(A) $\frac{\text{Actual effort}}{\text{Ideal effort}}$
(B) $\frac{\text{Ideal effort}}{\text{Actual effort}}$

(C) $\frac{100}{\text{Actual effort} \times \text{Ideal effort}}$

(D) $\frac{\text{Actual effort} \times \text{Ideal effort}}{100}$

40. In a lifting machine with efficiency 80% an effort of 20 kg is required to raise a load of 800 kg. The velocity ratio of machine is—
(A) 30 (B) 40
(C) 50 (D) 60

41. Mechanical advantage is the ratio between—
(A) Load and distance moved by the effort
(B) Load and distance moved by the load
(C) Load and effort
(D) Effort and load

42. In a machine, the ratio in between output and input is known as—
(A) Efficiency
(B) Mechanical advantage
(C) Velocity ratio
(D) Leverage

43. Generally in most of the machines, when mechanical advantage is greater than one, then its velocity ratio should also be—
(A) Less than one
(B) Equal to one
(C) Greater than one
(D) Equal to efficiency

44. If D and d are the diameter of wheel and axle respectively of a Weston's differential pulley block then the velocity ratio of the block will be—
(A) $\frac{D}{d}$ (B) $\frac{2D}{d}$
(C) $\frac{D}{D-d}$ (D) $\frac{2D}{D-d}$

45. At right angles to the direction of application of a force (F) the work done will be—
(A) Zero (B) Maximum
(C) Minimum (D) None of these

46. A structure which offers negligible or zero resistance on bending at any point is known as—
(A) Beam (B) Girder
(C) Arch (D) Cable or string

47. The shape which a cable takes (hanging freely) under a uniformly distributed load per metre run will be—
(A) Parabola (B) Hyperbola
(C) Straight line (D) Eliptical

48. The equation of a catenary in cartesian form is—
(A) $y = c \cosh x/c$ (B) $y = c \sinh x/c$
(C) $y = c \cosh^{-1} x/c$ (D) $y = c \sinh^{-1} x/c$

49. The most simplest geometrical form of a truss is—
(A) Triangle (B) Parallelogram
(C) Trapezium (D) Rhombus

50. A system of coplanar forces is in equilibrium when—
(A) Force polygon closes
(B) Funicular polygon closes
(C) Both force and funicular polygon close
(D) All the forces are coplanar concurrent

51. The condition of equilibrium for any system of forces in a plane is—
(A) The resultant couple must be zero
(B) The polygon of forces must close
(C) Both (A) and (B)
(D) None of these

52. If the number of members provided is more than the requirement, then the frame is called as—
(A) Perfect frame (B) Redundant frame
(C) Deficient frame (D) Portal frame

53. For a perfect frame, the relation between number of joints 'j' and number of members 'n' will be—
(A) $n = 2j$ (B) $n > 2j - 3$
(C) $n - 3 = 2j$ (D) $n = 2j - 3$

54. Compression members are generally called—
(A) Struts (B) Ties
(C) Purlins (D) Piers

55. The graphical method of determining forces in the members of truss is based on—
(A) Method of joints
(B) Method of section
(C) Both (A) and (B)
(D) None of these

56. Williot Mohr's diagram is used to determine deflection in—
(A) Beams (B) Trusses
(C) Rigid frames (D) All of them

57. The area under velocity-time graph gives—
(A) Acceleration
(B) Total distance travelled
(C) Velocity
(D) Speed

58. The uniform acceleration of an upward projected body will be—
(A) 9·81 m/sec (B) –9·81 m/sec
(C) 9·81 m/sec^2 (D) –9·81 m/sec^2

59. If a stone is thrown upward at an initial velocity 'u' then the maximum height reached by the stone will be—
(A) $\sqrt{2gh}$ (B) $2u/g$
(C) $\sqrt{\frac{u^2}{2g}}$ (D) $\frac{u^2}{2g}$

60. For a given velocity of a projectile the range is maximum when the angle of projection is—
(A) 30° (B) 45°
(C) 60° (D) 90°

61. When a particle is projected with an initial velocity u at an angle α with the horizontal, then the horizontal distance covered by the particle will be—
(A) $\frac{2u \sin \alpha}{g}$ (B) $\frac{u \sin \alpha}{g}$
(C) $\frac{u^2 \sin 2\alpha}{g}$ (D) $\frac{u^2 \sin \alpha}{2g}$

62. The unit of angular displacement is—
(A) Radian
(B) Radian per second
(C) Radian per second per second
(D) Pie (π)

63. The unit for acceleration is—
(A) Radian
(B) Radian per second
(C) Radian per second per second
(D) Pie (π)

64. Angular velocity is generally represented by—
(A) ω (B) π
(C) α (D) θ

65. The locus of all instantaneous centres is called as—
(A) Moment of inertia
(B) Centrode
(C) Centre of gravity
(D) Centre of pressure

66. The motion of the wheel of a bicycle is known as—
(A) Rotary
(B) Curvilinear
(C) Translatory
(D) Rotary and translatory

67. The number of oscillations per second is known as—
(A) Time period (B) Phase difference
(C) Frequency (D) Amplitude

68. The maximum displacement of a body, from its mean position is known as—
(A) Amplitude (B) Oscillations
(C) Beat (D) Frequency

69. The time taken by a particle for one complete oscillation is known as—
(A) Beat (B) Frequency
(C) Periodic time (D) Amplitude

70. In a simple pendulum of length l, the time period T is calculated from the formula—
(A) $\frac{\sqrt{l}}{g}$ (B) $2\pi\sqrt{\frac{g}{l}}$
(C) $2\pi\sqrt{\frac{l}{g}}$ (D) $\pi\sqrt{\frac{g}{l}}$

71. The time period of a second pendulum is—
(A) $\frac{1}{2}$ second (B) 1 second
(C) 2 second (D) 5 second

72. The period of oscillation of a simple pendulum depends upon—
(A) Mass of the bob
(B) Diameter of the bob
(C) Density of the string
(D) Length of the pendulum

73. The distance from the axis of reference where the whole mass or area of a body is assumed to be concentrated is known as—
(A) Lever arm
(B) Moment of inertia
(C) Centre of gravity
(D) Radius of gyration

74. In S.I. units the unit of work is—
(A) Kg-m (B) Newton
(C) Newton-m/sec (D) Newton-metre

75. In terms of work, power is defined as—
(A) Quantity of work
(B) Rate of doing work
(C) Capacity of doing work
(D) Rate of change of doing work

76. The unit of power generally adopted is—
(A) Kg-m/sec (B) Joule
(C) Newton-m/sec (D) Horse power

77. One horse power in metric system is equal to—
(A) 75 kg-metre/second
(B) 75 kg-m/hour
(C) 4500 kg-m/second
(D) 4500 kg-m/hour

78. In S.I. unit, the unit of power is—
(A) Joule (B) Watt
(C) Newton (D) Erg

79. In terms of work energy is defined as—
(A) Quantity of work
(B) Rate of doing work
(C) Capacity of doing work
(D) None of these

80. A body of mass '*m*' moving with a uniform velocity '*v*' then the energy stored by the body will be—
(A) $\frac{mv}{2g}$ (B) $\frac{mv^2}{g}$
(C) $\frac{mv^2}{g^2}$ (D) $\frac{mv^2}{2}$

81. The S.I. unit of kinetic energy is—
(A) Newton (B) Erg
(C) Joule (D) Kg-m/sec.

82. Due to the creep of belts, in a belt driven system there will be—
(A) Loss of motion
(B) Loss of power to be transmitted
(C) Extra movement of belt as compared to the follower
(D) All the above are correct

83. The member forces in a statically indeterminate truss—
(A) Can be obtained by graphic statics
(B) Can not be obtained by graphic statics
(C) May be obtained by graphic statics
(D) None of these

84. An ordinate in a funicular polygon represents—
(A) Shear force
(B) Resultant force
(C) Bending moment
(D) Equilibrium

85. The number of funicular polygons which can be drawn to pass through two specified points in the space diagram are—
(A) Zero (B) 1
(C) 2 (D) Infinity

86. A funicular polygon can not be made to pass through—
(A) One specified points
(B) Two specified points
(C) Three specified points
(D) More than three specified points

87. The condition of equilibrium for any system of forces in a plane is—
(A) That polygon of forces must close
(B) That resultant couple must be zero
(C) Both (A) and (B)
(D) None of these

88. Three forces are in equilibrium when—
(A) The force triangle closes
(B) The lines of action of three forces are concurrent
(C) Both (A) and (B)
(D) None of these

89. If forces of 1N, 2N, 3N, 4N, 5N and 6N act in order along the sides of a regular hexagon, then resultant is—
(A) 0 (B) 6N
(C) 12N (D) 21N

90. A system of coplanar forces acting on a rigid body can be reduced to—
(A) One force only
(B) One couple only

(C) One force and one couple only
(D) None of these

91. The product of a scalar and a vector is—
(A) Always a scalar
(B) Always a vector
(C) Neither a scalar nor a vector
(D) May be a scalar or may be a vector

92. Force polygon method is applicable for—
(A) Any coplanar force system
(B) A system of parallel forces only
(C) Concurrent coplanar forces only
(D) None of these

93. The force polygon representing a set of forces in equilibrium is a—
(A) Triangle (B) Open polygon
(C) Closed polygon (D) Parallelogram

94. The diagram showing the point of application and line of action of forces in their plane is called—
(A) Vector diagram
(B) Space diagram
(C) Force diagram
(D) Funicular diagram

95. The graphical method of determining the forces in the members of a truss is based on—
(A) Method of joint
(B) Method of section
(C) Both (A) and (B)
(D) None of these

96. If a body is acted upon by a number of coplanar non-concurrent forces it may—
(A) Rotate about itself without moving
(B) Move in any one direction
(C) Move in any direction rotating about itself
(D) All the above

97. The ratio of M.I. of a triangle and that of a rectangle having same base and height about their bases is—
(A) 1 (B) 2
(C) 3 (D) 4

98. The centre of gravity of a plane lamina will not be at its geometrical centre if it is a—
(A) Circle
(B) Right angled triangle
(C) Square
(D) Rectangle

99. Strain energy of any member may be defined as work done on it—
(A) To deform it
(B) To resist elongation
(C) To resist shortening
(D) All the above

100. The ratio of M.I. of a circular plate to that of a square plate for equal depth is—
(A) Less than one
(B) Equal to one
(C) Greater than one
(D) None of these

ANSWERS

1. (C)	2. (C)	3. (B)	4. (B)	5. (A)
6. (B)	7. (A)	8. (B)	9. (C)	10. (A)
11. (A)	12. (B)	13. (C)	14. (A)	15. (B)
16. (C)	17. (A)	18. (D)	19. (A)	20. (A)
21. (D)	22. (D)	23. (B)	24. (B)	25. (B)
26. (A)	27. (C)	28. (C)	29. (D)	30. (C)
31. (D)	32. (C)	33. (A)	34. (C)	35. (A)
36. (C)	37. (C)	38. (B)	39. (B)	40. (C)
41. (C)	42. (C)	43. (C)	44. (D)	45. (A)
46. (D)	47. (A)	48. (A)	49. (A)	50. (C)
51. (C)	52. (B)	53. (C)	54. (A)	55. (A)
56. (B)	57. (B)	58. (D)	59. (D)	60. (B)
61. (C)	62. (A)	63. (C)	64. (A)	65. (B)
66. (D)	67. (C)	68. (A)	69. (C)	70. (C)
71. (C)	72. (D)	73. (D)	74. (D)	75. (B)
76. (D)	77. (A)	78. (B)	79. (C)	80. (D)
81. (C)	82. (D)	83. (B)	84. (C)	85. (D)
86. (D)	87. (C)	88. (C)	89. (B)	90. (C)
91. (B)	92. (C)	93. (C)	94. (B)	95. (A)
96. (D)	97. (D)	98. (B)	99. (D)	100. (A)

Chapter-2

HYDRAULIC ENGINEERING

(1) Hydraulics

The science of fluids at rest is called hydrostatics while that of moving fluids or **hydrodynamics**. **Hydraulics** is the science which deals with the behaviour of fluids when subjected to a system of forces. It involves the study of kinematics, dynamics and statics.

(2) Fluid

The substances which flow are called fluids. Fluids include both liquids and gases.

(3) Viscosity

It is the property of a fluid which resists relative motion of its adjacent layers. It is primarily due to cohesion and molecular momentum exchange between fluid layers.

$$\text{Shearing stress } \tau \propto \frac{dv}{dy}$$

where, dv = Relative velocity between two adjacent fluid layers

dy = Distance measured normal to the direction of flow

$$\therefore \quad \frac{dv}{dy} = \text{Velocity gradient}$$

$$\tau \propto \mu \frac{dv}{dy}$$

where μ is known as the coefficient of viscosity. The unit of viscosity is poise.

$$1 \text{ poise} = \frac{1}{98{\cdot}1} \text{ kg-sec/m}^2$$

(4) Centre of pressure

The centre of pressure of a plane are in contact with a liquid is the point of the area at which the resultant thrust on one side of the area acts.

(5) Surface tension

The property of a liquid due to which its free surface tries to have minimum surface area and behaves as if it were under tension somewhat like a stretched elastic membrane is called surface tension. A small liquid drop has spherical shape, as due to surface tension the liquid surface tries to have minimum surface area and for a given volume, the sphere has minimum surface area.

(6) Pascal's law

The pressure at any point in a fluid at rest has the same magnitude in all directions. Thus when a certain pressure is applied at any point in a fluid at rest, the pressure is equally transmitted in all directions and to every point in the liquid.

(7) Laminar flow

The fluid particle move along smooth paths in laminars, or layers with one layer gliding smoothly over an adjacent layer.

(8) Turbulent flow

In this, the fluid particles move in very irregular paths, causing an exchange of momentum from one portion of fluid to another.

(9) Steady flow

Steady flow occurs when the fluid properties such as density, velocity and pressure at any given point do not change with the passage of time. Unsteady flow occurs when the fluid properties change with time.

(10) Uniform flow

It occurs when the flow characteristics at any given time remain the same at different points in the direction of flow. Non-uniform flow occurs when the flow characteristics at any time change with distance.

(11) Stream line

A stream line is an imaginary curve drawn through a flowing fluid in such a way that the tangent to it at any point gives the direction of the velocity of flow at that point. Thus there can be no component of the velocity at right angles to the stream line and hence there can be no flow of fluid across any stream line.

(12) Rotational flow

If the net rotation of fluid element is not zero, the flow is termed as rotational.

(13) Stream function

The flow rate or mass per unit time across any line between the two stream lines and per unit thickness is known as stream function. It is denoted by ψ.

Here, $$u = \frac{\partial\psi}{\partial y}, v = -\frac{\partial\psi}{\partial x}$$

$$\frac{\partial\psi}{\partial y}\cdot dy + \frac{\partial\psi}{\partial x}\cdot dx = 0$$

(14) Velocity potential

The velocity of flow in a certain direction will depend upon the potential difference which is known as velocity potential and denoted by ϕ.

Mathematically,

$$u = -\frac{\partial\phi}{\partial x}, \quad v = -\frac{\partial\phi}{\partial y}, \quad w = -\frac{\partial\phi}{\partial z}$$

where, u, v, w are the velocity component in x, y, z direction.

(15) Flow net

It is a graphical representation of stream lines and equipotential line which are perpendicular to each other.

The condition of orthogonality of stream lines and potential lines is

$$\frac{\partial\phi}{\partial x}\cdot\frac{\partial\psi}{\partial x} + \frac{\partial\phi}{\partial y}\cdot\frac{\partial\psi}{\partial y} = 0$$

Buoyancy and Archimedes' Principle

Accidentally Archimedes' discovered that when a body is immersed partly or wholly in a fluid, it is buoyed up with a force equal to the weight of the fluid displaced by the body. This principle is called Archimedes' principle.

When a body is partly or wholly dipped into a fluid, the fluid exerts force on the body due to hydrostatic pressure. At any small portion of the surface of the body, the force exerted by the fluid is perpendicular to the surface and is equal to the pressure at that point multiplied by the area. The resultant of all these constant forces is called up-thrust or buoyancy.

Bernoulli's theorem

According to it in case of steady flow of incompressible and non-viscous fluid through a tube of non-uniform cross-section, the sum of the pressure, the potential energy per unit volume and the kinetic energy per unit volume is same at every point in the tube.

i.e., $$P + \rho gh + \frac{1}{2}\rho v^2 = \text{Constant}$$

Applications of Bernoulli's principle :

(1) Action of atomiser

(2) Working of aeroplane

(3) Velocity of efflux

(4) Venturimeter.

Following are the dimensionless constants for a moving liquid—

(1) Reynold number

$$R_e = \frac{\text{Inertial force}}{\text{Viscous force}} = \frac{UL\rho}{\mu}$$

(2) Froude number

$$F_r = \sqrt{\frac{\text{Inertial force}}{\text{Gravity force}}} = \frac{U}{\sqrt{gl}}$$

(3) Euler number

$$E = \sqrt{\frac{\text{Inertial force}}{\text{Pressure force}}} = \frac{U}{\sqrt{\Delta P/\rho}}$$

(4) Drag coefficient

$$G_p = \frac{\text{External force}}{\text{Inertial force}} = \frac{F}{U^2L^2\rho/2}$$

(5) Mach number

$$M = \sqrt{\frac{\text{Inertial force}}{\text{Compressibility force}}} = \frac{U}{\sqrt{E/\rho}}$$

(6) Weber number

$$W = \frac{\text{Inertial force}}{\text{Surface tension force}}$$

$$= \frac{\rho U^2 l}{\sigma}$$

Flow through pipes

Pipe is a closed apparatus used for carrying fluids under pressure. The flow in the pipe may be laminar or turbulent. The losses of head in pipes are—

$$\text{Loss at entrance} = \frac{0{\cdot}5v^2}{2g}$$

$$\text{Loss at outlet} = \frac{v^2}{2g}$$

$$\text{Loss due to friction} = \frac{4fl\,v^2}{2gd}$$

Open channel flow—Flow in open channel or partly filled ducts necessarily applied only to liquids and is treated as incompressible flow. Laminar flow in open channels occurs only with very viscous liquid at very small velocities, or very shallow depths as a thin film.

Chezy's formula for discharge through an open channel.

$$Q = AC\sqrt{m_i}$$

Kutter's formula for Chezy's constant 'C'

$$C = \frac{23 + \dfrac{0{\cdot}00155}{S} + \dfrac{1}{N}}{1 + \left[23 + \dfrac{0{\cdot}00155}{S}\right] \times \dfrac{N}{\sqrt{m}}}$$

where, N = Kutter's constant.

Bazin's formula for Chezy's constant 'C'

$$C = \frac{157{\cdot}6}{1{\cdot}81 + \dfrac{R}{\sqrt{m}}}$$

Manning formula for mean velocity 'T'

$$v = \frac{1}{N} m^{2/3} S^{1/2}$$

In a circular channel.

Area of flow $A = r^2$

Specific energy of a flowing liquid.

$$E = h + \frac{v^2}{2g}$$

Economical Sections

Rectangular section

For maximum discharge the conditions are—

(i) $b = 2d$ and

(ii) Hydraulic mean depth $m = \dfrac{d}{2}$

Tapezoidal section

(i) $\dfrac{b + 2nd}{2} = d\sqrt{n^2 + 1}$

(ii) $m = \dfrac{d}{2}$

Specific energy

The sum of potential energy and kinetic energy at any section of a channel is called specific energy.

Hydraulic jump equations for rectangular channel are as follows—

(i) $\dfrac{y_2}{y_1} = \dfrac{1}{2}\left[\sqrt{1 + 8F_{r_1}^{\,2}} - 1\right]$

(ii) $h_L = \dfrac{(U_1 - U_2)^3}{2g\,(U_1 + U_2)} = \dfrac{(y_2 - y_1)^3}{4y_1y_2}$

(iii) $\dfrac{y_1}{y_2} = \dfrac{1}{2}\left[\sqrt{1 + 8F_{r_2}^{\,2}} - 1\right]$

(iv) $F_{r_2} = \dfrac{2\sqrt{2}\,F_{r_1}}{\left[\sqrt{1 + 8F_{r_1}^{\,2}} - 1\right]^{3/2}}$

Hydraulic Turbines

Reaction turbine

In reaction turbines the water enters the wheel under pressure and flows over the vanes. While passing over the vanes the pressure head is converted into velocity head and is finally reduced at atmospheric pressure before leaving the wheel.

Some of the reaction turbines are—

(1) Four neyron turbine

(2) Francis turbine

(3) Thomson turbine

(4) Keplan or propeller turbine

(5) Mixed flow turbine

Impulse turbines

In these turbines all the energy of water is converted into velocity before entering the wheel by expanding through a nozzle or guide vanes. These turbines can be placed at the foot of the fall and above the tail race.

Some of well known impulse turbines are—

(1) Pelton wheel

(2) Girard turbine

(3) Jonval turbine

(4) Banki turbine

(5) Turgo turbine

Both types of turbines are further subdivided on the basis of direction of flow of water through them.

(A) Radial flow—(i) Inward flow (ii) Outward flow

(B) Axial flow

(C) Mixed flow

Pumps

A pump is a mechanical device which when interposed in a pipeline transfers energy from some external source to the liquid flowing in a pipe. The types are—

(1) Reciprocating pump

(2) Centrifugal pump

Water power

A hydro power plant consists of the following works which form the nucleus of every installation—

(1) A structure to intercept and to regulate the free flow of water. It may be in the form of a dam or a weir or a barrage.

(2) Head race or intake conduit together with gates, trash rocks, surge tanks, *etc*., to control the entry of water from the reservoir on to the turbines.

(3) A power station where motive power obtained from water flow is utilised to generate electrical energy.

(4) Tail race or exist conduit to leads the water back into the river or artificial channel for its subsequent use for irrigation or other purposes.

The main component of hydro power plant could be as below—

(1) Force bay

(2) Intake arrangement

(3) Head race or intake conduit

(4) Surge tank

(5) Turbines and generators

(6) Power house

(7) Tail race

(8) Transformer lines

(9) Transmission lines

EXERCISE

1. Pressure at a point inside a liquid does not depend on—
(A) The depth of the point below the surface of the liquid
(B) The nature of the liquid
(C) The acceleration due to gravity at that point
(D) The shape of the containing vessel

2. Two stretched membranes of area 2 cm^2 and 3 cm^2 are placed in a liquid at the same depth. The ratio of the pressure on them is—
(A) 1 : 1 (B) 2 : 3
(C) 3 : 2 (D) $2^2 : 3^2$

3. Pressure is applied to an enclosed fluid. It is—
(A) Increased and applied to every part of the fluid
(B) Diminished and transmitted to the walls of the container
(C) Increased in proportion to the mass of the fluid and then transmitted
(D) Transmitted unchanged to every portion of the fluid and the walls of container

4. The principle of operation of Brahma's press is based on—
(A) Boyle's law (B) Pascal's law
(C) Dalton's law (D) Newton's law

5. As a bubble comes from the bottom of a lake to the top, its radius—
(A) Increases
(B) Decreases
(C) Does not change
(D) Becomes zero

6. When a bubble rises from the bottom of a lake to the surface, its radius doubles. The atmospheric pressure is equal to that of a column of water of height H. The depth of the lake is—

(A) H (B) 2H
(C) 7H (D) 8H

7. When a body is wholly or partially immersed in liquid it appears to lose weight. This loss of weight is equal to the weight of—

(A) Water displaced by the body
(B) Weight of liquid displaced by the body
(C) Equal volume of water
(D) Equal volume of liquid

8. Two pieces of metal when immersed in a liquid have equal upthrust on them, then—

(A) Both pieces must have equal weights
(B) Both pieces must have equal densities
(C) Both pieces must have equal volumes
(D) Both are floating to the same depth

9. If there were a smaller gravitational effect, which of the following forces do you think would alter in some respect ?

(A) Viscous force
(B) Archimedes uplift
(C) Electrostatic force
(D) Nuclear force

10. When a body is weighed in a liquid, the loss in its weight depends upon—

(A) Volume of the body
(B) Mass of the body
(C) Shape of the body
(D) CG of the body

11. Which of the following would a hydrogen balloon find easier to lift ?

(A) 1 kg of water
(B) 1 kg of steel
(C) 1 kg of light packed feather
(D) All the above

12. A hydrogen balloon released on the moon would—

(A) Climb up with an acceleration of 9·8 m/sec^2
(B) Climb up with an acceleration of 9·8 × 6 m/sec^2
(C) Neither climb nor fall
(D) Fall with an acceleration of 9·8/6 m/sec^2

13. An iron ball is weighed in air and then in water by a spring balance—

(A) Its weight in air is more than in water
(B) Its weight in water is more than in air
(C) Its weight is same both in air and water
(D) Its weight is zero in water

14. A body of weight W_1 displaces an amount of water W_2. When floating—

(A) $W_1 > W_2$ (B) $W_1 < W_2$
(C) $W_1 = W_2$ (D) A, B, C, any

15. When a ship floats on water—

(A) It displaces no water
(B) The weight of water displaced is equal to the weight of the ship
(C) The weight of water displaced is lesser than the weight of the ship
(D) The weight of water displaced is greater than the weight of the ship

16. For a body floating in water the apparent weight is equal to—

(A) Actual weight of the body
(B) Zero
(C) Weight of the body minus weight of the liquid
(D) None of the above

17. A wooden cylinder floats vertically in water with half of its length immersed. The density of wood is—

(A) Equal to that of water
(B) Half the density of water
(C) Double the density of water
(D) The question is incomplete

18. It is easier to swim in sea water than in river water because—

(A) Sea is wider than the river
(B) Sea is deeper than the river
(C) Density of sea water is greater than that of river
(D) Sea is having high waves all the time

19. When a loaded boat enters into the sea from a river, it rises because—
(A) There is more water in sea than in river
(B) Sea water is denser than river
(C) There is difference of temperature
(D) Sea is deeper than river

20. When a piece of ice floating in a beaker of water completely melts, the level of water in the beaker—
(A) Rises
(B) Falls
(C) Remains unchanged
(D) Rises or falls depending upon the quantity of ice and water

21. An ice cube containing a lead piece in it is floating in a glass of water. As ice melts the water level will—
(A) Fall
(B) Rise
(C) Remain stationary
(D) None of the above

22. A man is sitting in a boat which is floating in a pond. If the man drinks some water from the pond, the level of water in the pond will—
(A) Rise a little
(B) Sink a little
(C) Remain stationary
(D) None of the above

23. A boat with wood is floating in a lake. If the wood is thrown in the lake the water level will—
(A) Go up
(B) Go down
(C) Remain unchanged
(D) None of the above

24. A boat with scrap iron is floating in a lake. If the scrap iron is thrown in the lake the water level will—
(A) Go up
(B) Go down
(C) Remain unchanged
(D) None of the above

25. A wooden piece floats half submerged in a tub of water. If the system is transferred to a lift ascending with acceleration the wooden piece will—
(A) Sink a little more
(B) Rise a little
(C) Remain half submerged
(D) Will sink to the bottom

26. If in a satellite orbiting the earth, a cork is immersed in a jar of water and released, it will—
(A) Rise
(B) Sink
(C) Remain where left
(D) None of the above

27. A gale blows over a house. The force due to the gale on the roof is—
(A) Downward (B) Upward
(C) Horizontal (D) Zero

28. Action of a point-gun is based on—
(A) Bernoulli's principle
(B) Boyle's law
(C) Faraday's law
(D) Archimedes' principle

29. An aeroplane works on
(A) Archimedes' principle
(B) Pascal's law
(C) Bernoulli's principle
(D) Stoke's law

30. Magnus effect is very near to the—
(A) Magnetic field
(B) Electric field
(C) Bernoulli's theorem
(D) Magnetic effect of current

31. Liquids—
(A) Have no shape
(B) Can not be compressed
(C) Both (A) and (B)
(D) None of the above

32. The weight per unit volume of a liquid at a standard temperature and pressure is called—
(A) Specific weight (B) Mass density
(C) Specific gravity (D) None of these

33. The specific gravity of water at standard temperature (4°C) equals to—
(A) 0·999 (B) 1·000
(C) 1·010 (D) 0·500

34. The specific weight of water in S.I. units is—
(A) 9·81 gm (B) 981 gm
(C) 9·81 k N/m^3 (D) $9{\cdot}81 \times 10^3$ k N/m^2

35. As ideal fluid is—
(A) One which obeys Newton's law of viscosity
(B) One which satisfies continuity equation
(C) One which flows through pipes with least friction
(D) Frictionless and incompressible

36. An ideal flow of any fluid must fulfil the following—
(A) Newton's law of viscosity
(B) Newton's law of motion
(C) Boundary Layer theory
(D) Continuity equation

37. The variation in volume of a liquid with the variation of pressure is called its—
(A) Surface tension
(B) Compressibility
(C) Capillarity
(D) Viscosity

38. 1 atm. pressure is equivalent to—
(A) 1·315 kgf/cm^2 (B) 1000 newton
(C) 760 mm Hg (D) 14·7 pascal

39. The pressure intensity in kg/cm^2 at any point in a liquid is—
(A) W (B) Wh
(C) W/h (D) h /W

40. A barometer is used to measure—
(A) Very low pressure
(B) Very high pressure
(C) Pressure difference between two points
(D) Atmospheric pressure

41. Manometer is used to measure—
(A) Velocity of liquid
(B) Atmospheric pressure
(C) Pressure in pipe and channels
(D) Pressure in venturimeter

42. Pressure measured with the help of piezometer tube is—
(A) Atmospheric pressure
(B) Gauge pressure
(C) Absolute pressure
(D) Vacuum pressure

43. A body in neutral equilibrium will rotate about—
(A) Centre of gravity and metacentre
(B) Centre of pressure and metacentre
(C) Centre of gravity and centre of buoyancy
(D) Centre of gravity, centre of buoyancy and metacentre

44. The relationship $\frac{dp}{dx} = \frac{dz}{dy}$ is valid for—
(A) Irrotational flow
(B) Non-uniform flow
(C) Uniform flow
(D) Unsteady flow

45. The stream function for a two dimensional flow is given by $\psi = 2xy$. The velocity at (2, 2) is—
(A) $4\sqrt{2}$ (B) 4
(C) $2\sqrt{2}$ (D) $\sqrt{2}$

46. A cubical tank is filled with water. If side of tank is 1 m pressure on the base of the tank will be—
(A) 10000 kg (B) 5000 kg
(C) 1000 kg (D) 4000 kg

47. The point at which the resultant pressure on an immersed surface acts is known as—
(A) Centre of gravity
(B) Centre of depth
(C) Centre of pressure
(D) Centre of immersed surface

48. The buoyancy depends upon the—
(A) Pressure with which the liquid is displaced
(B) Weight of the liquid displaced
(C) Viscosity of the liquid
(D) Compressibility of the liquid

49. When the metacentre of a floating body is lower than the centre of gravity, then the body will be in—
(A) Unstable equilibrium
(B) Stable equilibrium
(C) Neutral equilibrium
(D) None of these

50. Bernoulli's theorem has been derived under the assumption that no external force acts on the liquid except—
(A) Atmospheric pressure
(B) Force due to gravity
(C) Wind pressure
(D) None of these

51. Bernoulli's theorem deals with the principle of conservation of—
(A) Energy (B) Momentum
(C) Mass (D) Force

52. In venturimeter, the ratio between throat diameter and pipe diameter is generally adopted as—
(A) 1 : 2 (B) 1 : 4
(C) 1 : 8 (D) 2 : 11

53. The Euler's equation for steady flow of an ideal fluid along a stream line is based on Newton's—
(A) First law of force
(B) Second law of motion
(C) Third law of momentum
(D) Law of friction

54. Bernoulli's equation is applied to—
(A) Venturimeter (B) Orifice meter
(C) Pitot tube (D) All of these

55. The velocity of the liquid flowting through the divergent portion of a venturimeter—
(A) Remains constant
(B) Increases
(C) Decreases
(D) Depends upon mass of liquid

56. Pitot tube is a device used in the flowing fluid for measuring—
(A) Discharge
(B) Pressure
(C) Velocity
(D) Kinetic and potential energy

57. Froude number is the ratio of inertia force to—
(A) Pressure force (B) Viscous force
(C) Gravity force (D) Elastic force

58. The ratio between inertia force and viscous force is called—
(A) Reynold's number
(B) Froude's number
(C) Weber's number
(D) Euler's number

59. Square root of the ratio of inertia force to the surface tension force is known as—
(A) Reynold's number
(B) Froude's number
(C) Weber's number
(D) Euler's number

60. Mach's number is square root of the ratio between inertia force and the—
(A) Viscous force
(B) Pressure force
(C) Elastic force
(D) Surface tension force

61. The total energy line lies over the hydraulic gradient line by an amount equal to—
(A) Velocity head (B) Pressure head
(C) Friction head (D) Datum head

62. The Reynold's number for pipe flow is given by—
(A) $\frac{VD}{\nu}$ (B) $\frac{VD\mu}{\delta}$
(C) $\frac{\rho VD}{\mu}$ (D) $\frac{VD}{\mu}$

63. For a turbulent flow, the Reynold's number should be more than—
(A) 4000 (B) 2700
(C) 2000 (D) 500

64. Flow in a pipe is laminar if the Reynold's number is—
(A) Less than 2000
(B) Between 2000 and 4000
(C) Between 4000 and 6000
(D) Equal to 10000

65. The difference between the theoretical and actual velocities of the jet at vena contracta is mainly due to—
(A) Eddy currents
(B) Velocity of approach
(C) Friction
(D) Boundary layers

66. At vena contracta jet has the minimum area of cross section and so the velocity of liquid at this section will be—
(A) Minimum (B) Maximum
(C) Average (D) Zero

67. The theoretical velocity of jet at vena contracta is—
(A) $2gh$ (B) $2g\sqrt{H}$
(C) $\sqrt{2gh}$ (D) $H\sqrt{2g}$

68. The basic equation which governs the motion of incompressible viscous fluid in laminar motion is called as—
(A) Hagen-Poiseullie equation
(B) Stokes equation
(C) Darcy-Weisbach equation
(D) Navier-Stokes equation

69. In a free vortex, velocity—
(A) Decreases with radius
(B) Increase with radius
(C) Is constant
(D) None of these

70. A piece of length more than double the diameter of orifice fitted externally or internally to the orifice is called a—
(A) Notch (B) Weir
(C) Mouth piece (D) Nozzle

71. If 'v' is the mean velocity of flow, then the inlet loss of energy will be—
(A) $\frac{v^2}{2g}$ (B) $\frac{2v^2}{2g}$
(C) $\frac{v^2}{4g}$ (D) $\frac{v^2}{g}$

72. The loss of energy at exit from a pipe having 'v' as mean velocity will be—
(A) $\frac{v^2}{2g}$ (B) $\frac{0 \cdot 5v^2}{2g}$
(C) $\frac{2g}{v^2}$ (D) None of these

73. In open channel water flows under—
(A) Force of gravity
(B) Atmospheric pressure
(C) Hydrostatic pressure
(D) Mechanical pressure

74. For a flow to be turbulent in an open channel Reynold's number should not be less than—
(A) 500 (B) 1000
(C) 1500 (D) 2000

75. When the depth of flow changes abruptly over a short distance, then the flow will be termed as—
(A) Rapidly varied flow
(B) Unsteady flow
(C) Gradually varied flow
(D) Turbulent flow

76. When the flow in open channel is gradually varied, the flow is said to be—
(A) Steady uniform flow
(B) Steady non-uniform flow
(C) Unsteady uniform flow
(D) Unsteady non-uniform flow

77. When Froude's number is more than 1, then the flow is termed as—
(A) Critical flow (B) Subcritical flow
(C) Streaming flow (D) Shooting flow

78. In an open channel flow the discharge corresponding to critical depth is—
(A) Minimum (B) Maximum
(C) Zero (D) Average

79. The separation of boundary layer takes place when the pressure gradient is—
(A) Negative (B) Positive
(C) Zero (D) Constant

80. If the Froude's number of flow in an open channel is more than 1·0 then the flow is said to be—
(A) Critical (B) Shooting
(C) Streaming (D) Transitional

81. Which one of the following conditions is a typical characteristic of critical flow ? (Symbols have their normal meanings)
(A) $\frac{Q^2T}{gA^3} = 1$ (B) $\frac{QT^2}{gA^3} = 1$
(C) $\frac{Q^2}{gA^3}$ (D) $\frac{Q^2T^2}{gA^2} = 1$

82. Which one of the following velocity fields represents a possible fluid flow ?
(A) $u = x, v = y$ (B) $u = x^2, v = y^2$
(C) $u = xy, v = x^2y^2$ (D) $u = x, v = -y$

83. The most economical section of a trapezoidal channel is one which has hydraulic mean depth equal to—
(A) $\frac{1}{2}$ depth (B) $\frac{1}{2}$ sloping side
(C) $\frac{1}{2}$ width (D) $\frac{1}{2}$(depth + width)

84. If weir is used as a spillway it will have a crest—
(A) Ogee shaped (B) Broad and flat
(C) Narrow (D) Cippolet

85. In a tank or channel, notch is provided to measure—
(A) Velocity (B) Pressure
(C) Discharge (D) Static energy

86. Weir may also be used to measure—
(A) Velocity of flow
(B) Pressure
(C) Discharge in a river or channel
(D) Kinetic energy

87. The sheet of water flowing through a notch or over a weir is known as—
(A) Nappe
(B) Crest
(C) Height of weir/notch
(D) Sill

88. The velocity head due to velocity of approach is given by the formula—
(A) $\frac{v^2}{2g}$ (B) $\frac{v^3 - u^3}{2g}$
(C) $\frac{v^2}{P^2}$ (D) $\frac{v^2}{P^2}\sqrt{2gH}$

89. In order to avoid cavitation the upstream edge of broad crested weir should have—
(A) Rounded corner (B) Sharp corner
(C) Parabolic corner (D) None of these

90. In a broad crested weir, width of crest of the weir will be—
(A) More than 0·5 H
(B) Less than 0·5 H
(C) Both (A) and (B)
(D) None of the two

91. In a gradually varied flow, if dy/dx is positive, then dE/dx will be—
(A) Zero
(B) Positive if $Y > Y_c$
(C) Negative if $Y > Y_c$
(D) None of these

92. The horizontal to vertical side slope in case of cippoletti weir ?
(A) 1 : 1 (B) 1 : 4
(C) 1 : 2 (D) 4 : 1

93. The concept of boundary layer was first introduced by—
(A) Newton (B) Prandtl
(C) Reynold (D) Kutter

94. The thickness of boundary layer at the leading (entrance) edge will be—
(A) Maximum (B) Minimum
(C) Zero (D) Average

95. Laminar sublayer is a thin boundary layer, formed just adjacent to the boundary in—
(A) Turbulent flow (B) Laminar flow
(C) Transition flow (D) Compressible

96. Across the boundary layer, pressure remains–
(A) More (B) Less
(C) Constant (D) Zero

97. As compared to laminar flow, in turbulent flow, boundary layers will be—
(A) Thinner (B) Thicker
(C) Same (D) Half

98. The tangential force exerted by the shearing stresses on an object submerged in a fluid is known as—
(A) Pressure drag (B) Surface drag
(C) Drag (D) Lift

99. Drag and lift forces acting on an object placed is a fluid stream are caused by—
(A) Pressure and viscosity
(B) Pressure and gravity
(C) Pressure and turbulence
(D) Viscosity and turbulence

100. The component of resultant fluid dynamic force, in the flow direction is called—
(A) Drag force (B) Lift force
(C) Viscous force (D) None of these

101. For a stream line body, the predominant drag force is—
(A) Friction drag force
(B) Pressure drag force
(C) Total drag force
(D) None of these

102. The pressure drag depends upon—
(A) The boundary layer formation
(B) The separation of boundary layer
(C) The size of wake
(D) Both (B) and (C) above

103. The friction drag is caused by—
(A) Separation of boundary layer
(B) Pressure distribution over the body surface
(C) Shear stress generated due to viscous action
(D) All of the above

104. When an ideal fluid flows past a sphere—
(A) Total drag is zero
(B) Minimum pressure occurs at rear stagnation point
(C) Minimum pressure occurs at front stagnation point
(D) None of the above

105. The friction drag is more than pressure drag for—
(A) A sphere
(B) A streamlined body
(C) A circular disc placed normal to the flow direction
(D) An air foil

106. Drag force is a function of—
(A) Velocity of the body
(B) Mass density of the fluid
(C) Projected area of the body
(D) All of the above

107. The ratio of friction drag to pressure drag is—
(A) 1/3 (B) 2/3
(C) 1 (D) 2

108. The bodies which have definite separation points fixed by their shapes are known as—
(A) Stream lined bodies
(B) Bluft bodies
(C) Both (A) and (B)
(D) None of the above

109. The force exerted by jet on vane is determined by—
(A) Energy conservation principle
(B) Momentum principle
(C) Continuity equation
(D) None of the above

110. The water jet after striking a stationary flat plate will be deflected at an angle of—
(A) 60° (B) 90°
(C) 110° (D) 135°

111. As compared to flat plate force of jet on a semi-circular vane will be—
(A) Half (B) Equal
(C) Double (D) Triple

112. In a semi-circular vane, the maximum efficiency will be—
(A) 0·50 (B) 0·592
(C) 0·667 (D) 0·817

113. Hydraulic turbines are the machine which convert water energy into—
(A) Mechanical energy
(B) Electrical energy
(C) Solar energy
(D) Automatic energy

114. The mass of fluid issued by the jet per second for striking the fixed plate will be—
(A) $\omega a V$ (B) $\frac{\omega a V}{g}$
(C) $\omega a Vg$ (D) ωVg

115. Pelton-wheel is a—
(A) Tangential flow turbine
(B) Radial flow turbine
(C) Mixed flow turbine
(D) Axial flow turbine

116. Specific speed of Pelton-wheels having single jet, varies between—
(A) 10 to 35 (B) 10 to 50
(C) 60 to 400 (D) 300 to 1000

117. Specific speed of Francis turbine varies in between—
(A) 10 to 35 (B) 10 to 50
(C) 60 to 400 (D) 300 to 1000

118. In a Pelton-wheel, hydraulic efficiency will be maximum when speed of the buckets as compared to velocity of the jet will be—
(A) Equal (B) Double
(C) Half (D) One fourth

119. The turbine suitable for low heads and high flow rates is—
(A) Pelton-wheel (B) Francis
(C) Keplan (D) All of these

120. The specific speed of a Francis turbine is in the range—
(A) 10 to 35 (B) 35 to 60
(C) 60 to 400 (D) 300 to 1200

121. Cavitation causes damage in turbine runner near the—
(A) Inlet on the concave side of blades
(B) Inlet on the convex side of blades
(C) Outlet on the convex side of blades
(D) Outlet on the concave side of blades

122. Two throw reciprocating pump means—
(A) Double cylindrical pump
(B) Double acting pump
(C) Duplex double acting pump
(D) Single acting pump

123. In a single acting reciprocating pump, water is delivered in—
(A) Suction stroke
(B) Delivery stroke
(C) Middle of suction stroke
(D) Middle of delivery stroke

124. Wicket gates are—
(A) Sluice gates of the dam
(B) Guide vanes of a reaction turbine
(C) Runner vanes of a reaction turbine
(D) Gate valve in a penstock

125. Specific speed is—
(A) Dimensionless
(B) Measured in rev/min.
(C) A speed
(D) A shape factor

126. The valve provided in a suction pipe is called—
(A) Float valve (B) Gate valve
(C) Foot valve (D) Two-way valve

127. Viscosity of an oil changes with—
(A) Temperature (B) Speed
(C) Weight (D) Type of bearing

128. The generation of hydroelectric power as compared to power generated by coal, oil etc. is —
(A) Cheaper (B) Costlier
(C) At par (D) Can not compare

129. The power channel that extends from the intake works to the power house is called—
(A) Lead race (B) Head race
(C) Diversion canal (D) Penstock

130. In a fluid flow the line of constant piezometric head passes through two points which have the same—
(A) Elevation (B) Pressure
(C) Velocity (D) Velocity potential

131. In a water wheel when water enters the buckets mounted on the periphery of a wheel at its top, then it will be known as—
(A) Over shot wheel
(B) Breast water wheel
(C) Under shot wheel
(D) Pelton-wheel

ANSWERS

1. (D)	2. (A)	3. (D)	4. (B)	5. (A)
6. (C)	7. (B)	8. (C)	9. (B)	10. (A)
11. (C)	12. (D)	13. (A)	14. (C)	15. (B)
16. (B)	17. (B)	18. (C)	19. (B)	20. (C)
21. (A)	22. (C)	23. (C)	24. (B)	25. (C)
26. (C)	27. (B)	28. (A)	29. (C)	30. (C)
31. (A)	32. (A)	33. (B)	34. (C)	35. (D)
36. (D)	37. (B)	38. (C)	39. (B)	40. (D)
41. (C)	42. (B)	43. (D)	44. (A)	45. (A)
46. (C)	47. (C)	48. (B)	49. (A)	50. (B)
51. (A)	52. (A)	53. (B)	54. (D)	55. (C)
56. (C)	57. (C)	58. (A)	59. (C)	60. (C)
61. (A)	62. (C)	63. (A)	64. (A)	65. (C)
66. (B)	67. (C)	68. (D)	69. (B)	70. (C)
71. (C)	72. (A)	73. (A)	74. (D)	75. (A)

76. (B)	77. (D)	78. (B)	79. (A)	80. (B)
81. (A)	82. (D)	83. (A)	84. (A)	85. (C)
86. (C)	87. (A)	88. (A)	89. (A)	90. (A)
91. (B)	92. (B)	93. (B)	94. (C)	95. (A)
96. (C)	97. (B)	98. (B)	99. (A)	100. (C)
101. (B)	102. (C)	103. (A)	104. (D)	105. (A)
106. (D)	107. (A)	108. (A)	109. (B)	110. (B)
111. (C)	112. (B)	113. (A)	114. (B)	115. (A)
116. (A)	117. (C)	118. (C)	119. (B)	120. (A)
121. (C)	122. (A)	123. (B)	124. (B)	125. (D)
126. (C)	127. (A)	128. (A)	129. (B)	130. (D)
131. (A)				

●●●

Chapter-3

STRENGTH OF MATERIALS

Properties of Materials

The following are the main properties of materials—

(1) Hardness

It is the property which enables the material to resist scratching, wear and abrasion, penetration by hard objects.

(2) Malleability

It is the property which enables the material to undergo great changes in size without rupture under compressive stress.

(3) Ductility

It is the property due to which material can be drawn into wires.

(4) Plasticity

The property of the material by virtue of which it can not return to its original position when the load is removed, is known as plasticity. In this case there is a permanent deformation in the material.

(5) Elasticity

It is the property of material by virtue of which material springs back to its original position with in certain limit (Elastic limit) with the removal of external load.

(6) Brittleness

It is the property of the material due to which it breaks under load without permanent distortion.

(7) Toughness

It is the property of the material enabling it to resist fracture due to high impact loads like hammer blows.

Testing of Materials

Different tests and methods used for testing materials are—

(i) **Tensile test machine**—It consists of a mean of applying load to specimen, a device to measure the applied load and accessories for griping the test piece, recording the load and power unit etc.

(ii) **Tensile test of ductile metals**

(iii) **Percentage elongation**

(iv) **Tensile test of brittle metals**

(v) **Compression test**—This test is opposite to tensile test with respect to the direction of the applied stress.

(vi) **Shear test**—A force acting parallel to the plane produces shear stress which is of two types : (a) Direct shear (b) Torsional shear.

(vii) **Hardness test**—This test is carried out by pressing a loaded diamond against the surface under test.

(viii) **Fatigue test**—This test is carried out in parts subjected to continuously varying stresses either in magnitude or from one kind stress to the reversed kind.

(ix) **Impact test**—This load applied suddenly on a machine component is called impact blow. The impact load may be applied in flexure, tension, compression or torsion. Test enables to determination of toughness.

Load

A load on a material is the combination of external forces acting on it. When it is constant then called **dead load** and when it is variable then called as **live load**.

Stress

When a body is loaded, the restoring force per unit area, comes into play inside the body is called stress. Its unit is Pascal.

$$\text{Stress} = \frac{\text{Force applied}}{\text{Area of cross-section}} = \frac{P}{A}$$

Strain

It is defined as the ratio of change in length, volume, or shape to the original length, volume or shape. It is dimensionless. It is of three types—

(i) Longitudinal strain $= \dfrac{\text{Change in length}}{\text{Original length}}$

(ii) Volume strain $= \dfrac{\text{Change in volume}}{\text{Original volume}}$

(iii) Shearing strain = Change in shear

Young's modulus

It is defined as the ratio of stress to longitudinal strain.

$$\text{Young's modulus (E)} = \frac{\text{Stress}}{\text{Strain}} = \frac{F/A}{\delta l/l}$$

Bulk modulus

It is defined as the ratio of stress to volume strain.

$$\text{Bulk modulus} = \frac{\text{Stress}}{\text{Strain}} = \frac{P}{\Delta v/V}$$

Modulus of rigidity

It is defined as the ratio of tangential stress to shearing strain.

Modulus of rigidity G or N

$$= \frac{\text{Stress}}{\text{Strain}} = \frac{F/A}{\theta}$$

Hooke's law

Under the elastic limit stress is directly proportional to strain.

$$\text{Stress} \propto \text{Strain}$$
$$\text{Stress} = \text{E strain}$$
$$E = \frac{\text{Stress}}{\text{Strain}}$$

where, E is a constant and called modulus of elasticity.

Poisson's ratio

The ratio of lateral strain to longitudinal strain is called Poisson's ratio and denoted by $\frac{1}{m}$

Relationship between elastic constants :

(i) $E = 3K\left(1 - \frac{2}{m}\right)$

(ii) $E = 2N\left(1 + \frac{1}{m}\right)$

(iii) $\frac{1}{E} = \frac{1}{9K} + \frac{1}{3N}$;

$E = \frac{9KN}{3K + N}$

Temperature stress

When expansion or contraction of a body due to a change in temperature is prevented thermal stress and thermal strain is produced.

Temperature stress $= E.\alpha.t$

where, E = Modulus of elasticity

α = Coefficient of thermal expansion

t = Change in temperature

Stress on inclined plane

Stress on a plane inclined through θ to the cross-section carrying a load P.

Normal stress $P_n = P\cos^2\theta$

Tangential stress $P_t = \dfrac{P\sin 2\theta}{2}$

Direct stress

When the section is subjected to axial pull or push acting normally across the section the stress so produced is direct stress.

Shear stress

The shear stress is produced when the body is subjected to two equal and opposite forces P and tends to slide two adjacent layers. The intensity of shear stress

$$q = \frac{P}{A}$$

Ultimate stress

The maximum stress which a material is subjected to before rupture is called ultimate stress.

Bending Moment

Beam

A structural member acted upon by a system of external forces at right angle to its axis is called beam.

The beams may be classified as—

(i) **Cantilever**—A beam whose one end is fixed and other end being free is called cantilever.

(ii) **Simply supported beam**—A beam whose both the ends are resting on supports, which may be walls or columns is called freely supported beam.

(iii) **Encastre beam**—A beam whose ends are built in into the walls is called encastre or built in beam.

(iv) **Continuous beam**—A beam which is continuous over intermediate supports is called continuous beam.

Simple bending

A beam or a cantilever is said to be subjected to simple bending when it bends under the action of uniform bending moment action of any load.

Assumptions in theory of simple bending

(i) Plane of transverse sections of the beam, remain plane even after bending.

(ii) The material of the beam always obeys Hooke's law and is assumed homogeneous.

(iii) The beam is initially straight of constant cross-section.

(iv) The moduli of elasticity for tension and compression are same.

(v) The stresses in the beam do not exceed elastic limit.

(vi) The beam bends to a circular arc.

(vii) The radius of curvature is large compared to the section dimensions.

(viii) The loads act perpendicular to the beam axis.

Mohr's theorem

(a) The angle in radians between the tangents to the elastic curve at two points on a straight member under bending is equal to the area between those two points.

(b) The deflection of a point on a straight member under bending in the direction perpendicular to the original straight axis of the member measured from the tangent at another point on the member, is equal to the member between those two points, where the deflection occurs.

Columns and struts

A member subjected to compressive stress is termed as post, strut or pillar. Main supporting members of building are generally called columns or pillars.

The load carried by short column is

$$P = p \times A$$

where, p is the intensity of stress and A is the area of cross-section of the column.

A column that fails primarily due to direct stress is called a short column and the one that fails primarily due to the bending stress is called the long column.

For short column, $\frac{l}{k} < 32$ or $l < 8\phi$

where, ϕ = Diameter

For medium column,

$$\frac{l}{k} < 120 > 32$$

or $l < 30\psi < 8\phi$

For long column, $\frac{l}{k} > 120$ or $l > 30\phi$

Euler's Column Formulae

(i) Both ends hinged

$$W = \frac{\pi^2 EI}{l^2}$$

(ii) One end fixed and the other end free

$$W = \frac{\pi^2 EI}{4l^2}$$

(iii) Both ends fixed

$$W = \frac{4\pi^2 EI}{l^2}$$

(iv) One end fixed, other end hinged

$$W = \frac{2\pi^2 EI}{l^2}$$

Equivalent lengths for various end conditions—

S. No.	Type (where L = Actual length)	Effective length of member l
1.	Effectively held in position and restrained in direction at both ends	0·67 L
2.	Effectively held in position at both ends and restrained in direction at one end.	0·85 L
3.	Effectively held in position at both ends but not restrained in direction.	L
4.	Effectively held in position and restrained in direction at one end and at the other end effectively restrained in direction but not held in position.	L
5.	Effectively held in position and restrained in direction at one end and at the other end partially restrained in direction but not held in position.	1·5 L
6.	Effectively held in position and restrained in direction at one end but not held in position or restrained in direction at the other end	2·0 L

Rankine's formula

For long column with both ends hinged carrying a load W is given by

$$W = \frac{f_c.A}{1 + a\left(\frac{l}{k}\right)^2}$$

where, f_c is crushing stress

k = Least value of radius of gyration

a = Rankine's constant.

Springs

Spring is a device in which the material is arranged in such a way that it can undergo considerable change without being permanently distorted. Spring are of three types—

(1) Laminated spring

(2) Closely coiled helical spring

(3) Open coiled helical spring

Thin walled cylinder

Thin walled cylindrical vessels are used to store fluids under pressure upto less than 2500 kg/cm^2. Beyond this pressure vessels used in a way. The thickness of thin walled cylinder is about $\frac{d}{20}$ cm.

Due to internal pressure in the thin walled cylinder the three types of stresses may be observed—

(i) Hoop or circumferential stress acting across the longitudinal section.

(ii) Longitudinal stress acting across the section perpendicular to the longitudinal section.

(iii) Radial stress acting radially, being too small and can be ignored.

These three types of stresses are mutually perpendicular to the area.

Thick walled cylinders

Thick wall cylindrical vessels are used to store fluids under pressure more than 2500 kg/cm^2. A thick walled cylinder is subjected to the following three principal stresses—

(i) The radial pressure p

(ii) The hoop stress sf_n

(iii) The longitudinal tensile stress P

Shear centre

It is the point of intersection of bending axis and the plane of transverse section of a beam. The beams composed of thin sections generally are rolled I-sections, angles, channels etc. Therefore load must act through the shear centre to avoid twisting of the beam.

Shear flow

The shear flow at a point in the section is the force per unit length, and is expressed by

$$\tau t = \frac{VQ}{I}$$

where,

τ = Shear stress

t = Thickness

V = Total vertical load

Q = $a\,\bar{y}$ analogous

I = M.O.I. of cross-section about N.A.

EXERCISE

1. Which of the following is most elastic in the language of engineering in terms of strength of material ?
   (A) Air (B) Hydrogen
   (C) Water (D) Mercury

2. Which of the following substances has the highest value of the Young's modulus ?
   (A) Steel (B) Rubber
   (C) Wood (D) Plastic

3. The given diagram shows the change x in the length of a thin uniform wire caused by the application of stress F at two different temperatures T_1 and T_2. The variation suggest that—

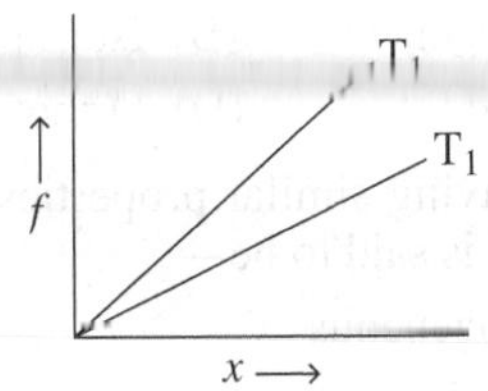

   (A) $T_1 > T_2$ (B) $T_1 < T_2$
   (C) $T_1 = T_2$ (D) None of these

4. Dimensional formula for modulus of elasticity is—
   (A) $[MLT^{-2}]$ (B) $[ML^{-1}T^{-2}]$
   (C) $[ML^{-2}T^{-1}]$ (D) $[ML^{-2}T^{-2}]$

5. The Young's modulus of a wire is numerically equal to the stress which will—
   (A) Not change the length of the wire
   (B) Double the length of the wire
   (C) Increase the length by 50%
   (D) Change the radius of the wire to half

6. A wire can be broken by applying a load of 20 kg weight. The force required to break the wire of twice the diameter is—
   (A) 20 kg weight
   (B) 5 kg weight
   (C) 50 kg weight
   (D) 160 kg weight

7. The following four wires are made of the same material. Which of these will have the largest extension when the same tension is applied ?
   (A) Length = 50 cm and diameter = 0·5 mm
   (B) Length = 100 cm and diameter = 0·5 mm
   (C) Length = 200 cm and diameter = 2 mm
   (D) Length = 300 cm and diameter = 3 mm

8. One end of a uniform wire of length L and of weight W is attached rigidly to a point in the roof and a weight W_1 is suspended from its lower end. If S is the area of cross-section of the wire, the stress in the wire at a height (3L/4) from its lower end is—
   (A) W_1/S
   (B) $(W_1 + W/4)/S$
   (C) $[W_1 + (3W/4)]/S$
   (D) $(W_1 + W)/S$

9. Energy stored per unit volume in a stretched wire is—
   (A) $\frac{1}{2}$ Load × Strain
   (B) Load × Strain
   (C) Stress × Strain
   (D) $\frac{1}{2}$ Stress × Strain

10. For gases—
    (A) Isothermal elasticity is greater than adiabatic elasticity
    (B) Adiabatic elasticity is greater than isothermal elasticity
    (C) Both the elasticities are equal
    (D) None of these

11. A given quantity of an ideal gas is at pressure P and absolute temperature T. The isothermal bulk modulus of the gas is—
    (A) $\frac{2}{3}$P (B) P
    (C) $\frac{3}{2}$P (D) 2P

12. The isothermal bulk modulus of a perfect gas at pressure P is equal to—
    (A) P (B) P/γ
    (C) $P\gamma$ (D) P^{γ}

13. The isothermal bulk modulus of a perfect gas at atmospheric pressure is—
(A) $1{\cdot}03 \times 10^5$ N/m^2
(B) $1{\cdot}03 \times 10^6$ N/m^2
(C) $1{\cdot}03 \times 10^{10}$ N/m^2
(D) $1{\cdot}03 \times 10^{11}$ N/m^2

14. A beam of metal supported at the two end is loaded at the centre. The depression at the centre is proportional to—
(A) Y^2 (B) Y
(C) 1/Y (D) $1/Y^2$

15. The upper end of a wire of radius 4 mm and length 100 cm is clamped and its other end is twisted through an angle of 30°, the angle of shear is—
(A) 12° (B) 0·12°
(C) 1·2° (D) 0·012°

16. The Poisson's ratio cannot have the value—
(A) 0·7 (B) 0·2
(C) 0·1 (D) 0·5

17. What is the relation between Y, β and η for isotropic material ?
(A) $\eta = \frac{3\beta Y}{9\beta + Y}$ (B) $\eta = \frac{9\beta\eta}{4\beta + Y}$
(C) $\eta = \frac{9\beta Y}{9\beta - Y}$ (D) $Y = \frac{9\beta\eta}{3\beta + \eta}$

18. The ability of a material to resist deformation is said to be—
(A) Strength (B) Toughness
(C) Plasticity (D) Brittleness

19. Ability of a material to deform permanently under compression without breaking is called—
(A) Malleability (B) Ductility
(C) Softness (D) Hardness

20. The ability of a material to deform without breaking is called—
(A) Resilience (B) Creep
(C) Plasticity (D) Elasticity

21. The steel used for cutting tools is known as—
(A) Mild steel
(B) High carbon steel
(C) Medium carbon steel
(D) None of these

22. The strength of cast iron casting as compared to cast steel casting is—
(A) More (B) Less
(C) Equal (D) None of these

23. The corrosion resistance of a material is its—
(A) Optical property
(B) Chemical property
(C) Thermal property
(D) Physical property

24. The property of a material enabling it to resist deformation under stress is called—
(A) Toughness (B) Stiffness
(C) Strength (D) Brittleness

25. If a material has identical properties in all directions, it is said to be—
(A) Homogeneous (B) Isotropic
(C) Elastic (D) Orthotropic

26. A body having similar properties throughout its volume is said to be—
(A) Homogeneous
(B) Isotropic
(C) Continuous
(D) Anisotropic

27. The unit of stress in S. I. units is—
(A) kg/cm^2 (B) kg/mm^2
(C) N/mm^2 (D) N/cm^2

28. The unit of stress in M.K.S. units is—
(A) kg/cm^2 (B) kg/cm
(C) N/m^2 (D) kg × m

29. Under external loads the stress is developed in the body due to—
(A) Heating
(B) Deformation of particles
(C) Cooling of particles
(D) None of the above

30. The deformation per unit length is called—
(A) Tensile stress (B) Compressive stress
(C) Shear stress (D) Strain

31. The unit of strain is—
(A) No unit (B) Millimetre
(C) Kg/cm^3 (D) Kg/cm^2

32. The ratio of linear stress to the linear strain is called—
(A) Modulus of rigidity
(B) Modulus of elasticity
(C) Bulk modulus
(D) Poisson's ratio

33. The greater stress to which a material can be subjected with any permanent deformation is said to be—
(A) Limit of proportionality
(B) Limit of elasticity
(C) Yield stress
(D) Limit of plasticity

34. The compression test is mainly done for testing—
(A) Brittleness of the material
(B) Ductility of the material
(C) Plasticity of the material
(D) Porosity of the material

35. The impact test assess the property of material such as its—
(A) Hardness (B) Strength
(C) Toughness (D) Brittleness

36. For fatigue test the pieces are tested—
(A) Upto the load limit
(B) Upto the line limit
(C) Upto some time of cycles
(D) Till fracture

37. Which of the following materials is more elastic ?
(A) Steel (B) Plastic
(C) Wood (D) Rubber

38. When a steel rod is pulled from opposite ends, the stress developed will be—
(A) Compressive stress
(B) Shear stress
(C) Bending stress
(D) Tensile stress

39. A live load is—
(A) One that occurs due to dead bodies
(B) One that does not exist actually
(C) One that remains constant
(D) One that does not remain constant

40. A live load—
(A) Exists only when required
(B) Is actually applied by putting weights
(C) Is one that varies continuously
(D) Is mainly due to self weight of the object

41. Modulus of elasticity is—
(A) $\frac{\text{Compressive stress}}{\text{Tensile strain}}$
(B) $\frac{\text{Tensile stress}}{\text{Compressive strain}}$
(C) $\frac{\text{Compressive stress}}{\text{Compressive strain}}$
(D) $\frac{\text{Tensile stress}}{\text{Compressive or tensile strain}}$

42. The modulus of elasticity is expected to have least value for—
(A) Aluminium (B) Glass
(C) Copper (D) Wood

43. For non-dilatant materials, the maximum value of Poisson's ratio is—
(A) 0·25 (B) 0·55
(C) 0·67 (D) 0·90

44. Which of the following has the highest value of Poisson' ratio—
(A) Rubber (B) Steel
(C) Aluminium (D) Copper

45. Hooke's law is valid—
(A) Only above elastic limit
(B) Only within elastic limit
(C) Only within plastic limit
(D) Till a substance breaks under load

46. Which of the following is the unit of Young's modulus of elasticity ?
(A) Newton/sq. cm. (B) Pounds/cu. m.
(C) Kg/cm (D) Dynes/ sq. cm

47. The modulus of elasticity for copper is more than that of lead. It can be, therefore, concluded that copper is—
(A) More elastic (B) More plastic
(C) Less elastic (D) Less plastic

48. When a bucket is pulled up by a string, the stress produced in the string is—
(A) Compressive (B) Tensile
(C) Shear (D) Torsional

49. When a block of wood is pushed, the stress produced in it will be—
(A) Shear (B) Frictional
(C) Compressive (D) Tensile

50. Poisson's ratio is the same as—
(A) $\frac{\text{Linear strain}}{\text{Lateral strain}}$
(B) $\frac{\text{Lateral strain}}{\text{Linear strain}}$
(C) $\frac{1}{\text{Linear strain} \times \text{Lateral strain}}$
(D) Linear strain × Lateral strain

51. The Poisson's ratio for steel varies from—
(A) 0·23 to 0·27 (B) 0·25 to 0·33
(C) 0·31 to 0·34 (D) 0·32 to 0·42

52. The Poisson's ratio for cast iron varies from—
(A) 0·23 to 0·27 (B) 0·25 to 0·33
(C) 0·31 to 0·34 (D) 0·32 to 0·42

53. The ratio of stress to volumetric strain is known as—
(A) Bulk modulus
(B) Young's modulus
(C) Modulus of rigidity
(D) Poisson's ratio

54. Hooke's law holds good upto—
(A) Elastic limit (B) Yield limit
(C) Breaking point (D) Plastic limit

55. The working stress in a material should be—
(A) Equal to its ultimate stress
(B) Equal to its lower yield point-stress
(C) Less than its yield point stress
(D) More than its yield point stress

56. For which of the following materials, the Poisson's ratio is expected to the least ?
(A) Steel (B) Cast iron
(C) Concrete (D) Copper

57. The ratio of the Young's modulus to the bulk modulus is—
(A) More than one third
(B) Less than one third
(C) Less than one half
(D) None of these

58. The bending moment on a section is maximum where shear force is—
(A) Zero (B) Minimum
(C) Maximum (D) Changing sign

59. The shear force on a beam—
(A) Causes linear transverse development
(B) Causes separation of fibres
(C) Causes rupture in the beam section
(D) Results in rotation of the planes

60. Shear froce (V) and bending moment (M) are related by—
(A) $V = \frac{M}{Z}$ (B) $V = \frac{dM}{dx}$
(C) $V = \int M dx$ (D) $V = \int MZ dx$

61. The shear force is generally associated with—
(A) Normal thrust
(B) Bending moment
(C) Rupture of beam
(D) None of the above

62. The strength of a beam section depends upon—
(A) Its sectional area
(B) Its sectional modulus
(C) Distance of its base from N. A.
(D) Its length

63. The point of contraflexure occurs in case of—
(A) Cantilevers only
(B) Overhanging beams only
(C) Continuous beams
(D) All types of beams

64. If shear force is zero along a section, the bending moment at that section will be—
(A) Zero
(B) Minimum
(C) Maximum
(D) Either minimum or maximum

65. A reinforced concrete beam is considered as made of—
(A) Homogeneous material
(B) Heterogeneous material
(C) Composite material
(D) Isotropic material

66. In a simple supported beam, along the neutral axis—
(A) Fibres do not undergo strain
(B) Fibres get twisted
(C) Fibres undergo maximum strain
(D) Fibres undergo minimum strain

67. On a simply supported beam, the shear force is proportional to—
(A) Algebraic sum of the forces
(B) Sum of transverse forces
(C) Displacement of the neutral axis
(D) None of the above

68. The point in a beam where the shear force changes sign is called the point of—
(A) Max. B. M.
(B) Zero shear
(C) Zero curvature
(D) Maximum curvature

69. A beam extending beyond the supports is called—
(A) Simply supported beam
(B) Cantilever beam
(C) Overhanging beam
(D) Continuous beam

70. A cantilever beam is one which is—
(A) Fixed at both ends
(B) Fixed at one end and free at other end
(C) Supported at its ends
(D) Supported at more than two points

71. The maximum bending moment of a cantilever beam lies at—
(A) The free end
(B) The fixed end
(C) 1/4 from free end
(D) 1/4 from fixed end

72. The bending moment diagram for a cantilever carrying uniformly distributed load will be—
(A) A triangle
(B) A parabola
(C) A cubic parabola
(D) A rectangle

73. The bending moment diagram for a cantilever beam subjected to bending moment at the free end will be—
(A) A triangle (B) A rectangle
(C) A parabola (D) A cubic parabola

74. Inclined loads generally cause—
(A) Thrust in the beam
(B) Twisting of the beam
(C) Points of contraflexure
(D) All the above

75. The expression $\Sigma y^2 \delta a$ is called the—
(A) Moment of inertia
(B) Bending moment
(C) Deflection
(D) Torsion

76. Out of the following mild steel sections, the most economical section is—
(A) 1-section
(B) Circular section
(C) Rectangular section
(D) Channel-section

77. The deflection of the beam may be reduced by—
(A) Decreasing the depth of beam
(B) Increasing the span
(C) Providing greater end restrains
(D) Any of above methods

78. A continuous beam is one which is—
(A) Fixed at both ends
(B) Fixed at one end and free at other end
(C) Supported on more than two supports
(D) Extending beyond the supports

79. The ratio of maximum deflection of a beam to its span is called—
(A) Section modulus of the beam
(B) Modulus rigidity of the beam
(C) Stiffness of the beam
(D) Bulk modulus of the beam

80. The product E. I. is also known as—
(A) Polar moment
(B) Flexual rigidity
(C) Stiffness
(D) Modulus of rigidity

81. The maximum deflection of a beam of length (L) simply supported at its ends with a central load W will be—
(A) $\frac{WL}{48 EI}$ (B) $\frac{WL^2}{48 EI}$
(C) $\frac{W^2 L}{48 EI}$ (D) $\frac{WL^3}{48 EI}$

82. An imaginary beam for which the load diagram is $\frac{M}{EI}$, the diagram will be known as—
(A) Simply supported beam
(B) Conjugate beam
(C) Laminated beam
(D) Fixed beam

83. In a fixed beam having a uniformly distributed load over the whole span, the moments will be calculated by the formula—
(A) $\frac{WL^2}{12}$ (B) $\frac{WL^3}{8}$
(C) $\frac{WL^2}{8}$ (D) $\frac{WL^3}{12}$

84. The ratio of average shear stress to maximum shear stress for a circular section is—
(A) 2 (B) 2/3
(C) 3/2 (D) 3/4

85. The maximum shear strain occurs on—
(A) Principal planes
(B) 45° with principal planes
(C) 90° with principal planes
(D) Independent of the principal planes

86. The shear stress on principal plane is—
(A) Maximum (B) Minimum
(C) Zero (D) None of these

87. The maximum shear stress is equal to the—
(A) Sum of the principal stresses
(B) Difference of the principal stresses
(C) Half of the difference of the principal stresses
(D) Half of the sum of the principal stresses

88. The sum of the normal stresses is—
(A) Variable
(B) Invariant
(C) Depends on planes
(D) None of these

89. The radius of Mohr's circle for two unlike principal stresses of magnitude p is—
(A) p (B) $p/2$
(C) $p/4$ (D) None of these

90. The plane which is not subjected to shear stress (tangential stress) is known as—
(A) Compound plane
(B) Principal plane
(C) Simple plane
(D) Non-shear plane

91. The determination of stresses on oblique sections of a strained material is an important step in the design of—
(A) Modern framed structures
(B) Riveted connections
(C) Welded connections
(D) None of these

92. Strain energy is given by—
(A) $\frac{1}{2}$ × Stress × Volume
(B) $\frac{1}{2}$ × Stress × Strain × Volume
(C) $\frac{1}{2}$ × Stress × Strain
(D) None of these

93. Strain energy of a member may be defined as work done on it—
(A) To deform it
(B) To resist elongation
(C) To resist shortening
(D) All the above

94. The basic perfect frame is a—
(A) Triangle (B) Rectangle
(C) Hexagon (D) Square

95. The S. I. unit of torque is—
(A) Kg cm (B) Newton-metre
(C) Kg/cm^2 (D) Kg-cm

96. The shear stress intensity changes from centre to the surface of the shaft with—
(A) Uniform rate
(B) Varying rate
(C) Remains same
(D) All the above

97. The shear stress intensity is maximum at—
   (A) Axis of the shaft
   (B) Surface of the shaft
   (C) Any inside layer of the shaft
   (D) All the above

98. The effective length of a chimney fixed in the ground, as compared to its actual length will be—
   (A) $L = 2l$ (B) $L = l/2$
   (C) $L = l$ (D) $L = \sqrt{2}\, l$

99. Euler's formula is not valid for mild steel column when slenderness ratio is—
   (A) More than 80 (B) More than 120
   (C) Less than 80 (D) More than 30

100. Rankine's formula is generally adopted when slenderness ratio lies in between—
   (A) 0 140 (B) 0 100
   (C) Less than 80 (D) Any value

101. Cylinders are said to be thin if their wall thickness '*t*' is given by—
   (A) $t = \frac{d}{20}$ (B) $t > \frac{d}{20}$
   (C) $t < \frac{d}{20}$ (D) $t = \frac{d}{20}$

102. Thick walled cylinders are supported to resist the pressure above—
   (A) 1000 kg/cm^2 (B) 2000 kg/cm^2
   (C) 2500 kg/cm^2 (D) None of these

103. The hoop stress induced in a thick cylinder due to radial pressure will be—
   (A) Tensile (B) Compressive
   (C) Shear (D) Bond

104. In a compound cylinder, hoop stress throughout the metal is—
   (A) Uniform (B) Variable
   (C) Zero (D) Maximum

105. In a particular material, if the modulus of rigidity is equal to the bulk modulus, then the Poisson's ratio will be—
   (A) $\frac{1}{8}$ (B) $\frac{1}{4}$
   (C) $\frac{1}{2}$ (D) 1

106. The rate of change of bending moment is equal to—
   (A) Shear force
   (B) Deflection
   (C) Slope
   (D) Poisson's ratio × Deflection

ANSWERS

1. (D)	2. (A)	3. (A)	4. (B)	5. (B)
6. (C)	7. (A)	8. (C)	9. (D)	10. (B)
11. (D)	12. (A)	13. (A)	14. (C)	15. (D)
16. (A)	17. (D)	18. (A)	19. (A)	20. (C)
21. (B)	22. (B)	23. (B)	24. (B)	25. (B)
26. (A)	27. (C)	28. (A)	29. (B)	30. (D)
31. (A)	32. (B)	33. (C)	34. (A)	35. (B)
36. (D)	37. (A)	38. (D)	39. (C)	40. (C)
41. (C)	42. (D)	43. (B)	44. (A)	45. (B)
46. (A)	47. (A)	48. (B)	49. (C)	50. (B)
51. (A)	52. (B)	53. (A)	54. (A)	55. (C)
56. (C)	57. (A)	58. (D)	59. (A)	60. (B)
61. (B)	62. (B)	63. (B)	64. (D)	65. (B)
66. (A)	67. (B)	68. (A)	69. (C)	70. (B)
71. (B)	72. (B)	73. (B)	74. (A)	75. (A)
76. (A)	77. (C)	78. (C)	79. (C)	80. (B)
81. (A)	82. (B)	83. (A)	84. (D)	85. (B)
86. (C)	87. (C)	88. (B)	89. (A)	90. (B)
91. (A)	92. (B)	93. (B)	94. (A)	95. (B)
96. (A)	97. (A)	98. (C)	99. (C)	100. (D)
101. (C)	102. (C)	103. (A)	104. (A)	105. (A)
106. (A)				

Chapter-4

SURVEYING

Surveying

It is the science of determining relative positions of objects on the surface of the earth by taking measurements of distances, directions, and elevations and plotting them to convenient reduced size on papers.

Basic principles

(i) Locating a point on the surface of the earth by at least two reference points.

(ii) Working from whole to the part. In this system first a system of control points are fixed with great precision.

Plane surveying

The survey where the effect of curvature of the earth is neglected assuming the earth's surface to be plane is called plane surveying. Generally areas less than 260 sq. km are treated as plane. For engineering works, this type of survey is generally followed.

Geodetic surveying

That survey, where the curvature of the earth is taken into account is called geodetic surveying. It is known as trigonometrical survey.

Scale

Scale is the fixed ratio that a distance on map bears with the corresponding distance on the ground *i.e.*, 1 cm = 10 m.

Representative fraction (R.F.)

The ratio of map distance to the corresponding ground distance is called representative fraction. It is independent of unit of measurement.

Chain survey

It is the method of surveying in which the area is divided into a network of triangles and the sides of the various triangles are measured directly in the field with chain or tape and no angular measurement are taken.

Station

The end points of a chain line are called station and the station on beginning or end of main chain line is known as main survey station. The tie station are selected anywhere on the chain line.

Base line

The longest line of the main survey lines is designated as the base line.

Tie line

The line connects tie stations or subsidiary stations on the main survey lines and is provided primarily with the object to facilitate taking offsets to objects distant from the main lines.

Offset

It is the lateral distance of an object or ground feature measured from a survey line. There are two types of offsets—

(i) Perpendicular offset

(ii) Oblique offset

Various instruments are used in chain survey. They are—

(1) Chain (2) Arrows (3) Pegs (4) Ranging rods (5) Offset rods (6) Plasterer's laths and whites (7) Plumb bob

The following are the precise instruments used for setting out right angles—

(1) Cross staff

(2) French cross staff

(3) Adjustable cross staff

(4) Optical square

(5) Prism square

Survey chain

Various chains used for the survey work are—

(i) **Metric chains**—They are available in lengths of 5, 10, 20 and 30 metre.

(ii) **Gunless chain**—They are available in 66 ft. length provided with 100 links.

(iii) **Engineer's chain**—It is 100 ft. long with 100 links.

(iv) **Revenue chain**—It is 33 ft. long provided with 10 links.

Tapes

These are used for more accurate measurement and are classified according to the material of which they are made.

For example :

(i) Cloth or linen tape

(ii) Metallic tape

(iii) Steel tape

(iv) Invar tape

Errors

Errors may arise due to any of the following reasons—

(i) **Instrumental**—Imperfect or faulty adjustments of instruments cause instrumental error.

(ii) **Personal**—Errors due to lack of perfection of human sight in observing are called personal errors.

(iii) **Natural**—Errors due to variations in natural phenomena such as temperature, humidity, refraction, magnetic declination etc. are called natural errors.

Errors in survey work may be classification as :

(i) **Mistakes**—These errors arise from inattention, inexperience, carelessness and mis-judgement or confusion in the mind of an observer.

(ii) **Cumulative errors**—These are the errors which under the same conditions, will always be of the same size and sign.

(iii) **Accidental or compensation errors**—These are the errors which remain even after mistakes and cumulative errors have been eliminated and are caused by a combination of reasons over which the observer has no control.

Corrections in linear measurements

For precise measurements, the following corrections are generally applied—

(i) Correction for standard length

(ii) Correction for alignment

(iii) Correction for slope

(iv) Correction for tension

(v) Correction for temperature

(vi) Correction for sag.

Positive and negative errors

Positive errors are those which make the result too great and negative errors make result too small.

	Errors	Nature of Error
1.	Erroneous length of chain	Cumulative, positive /negative
2.	Bad ranging	Cumulative, positive
3.	Bad straightening	Cumulative, positive
4.	Careless holding / marking	Compensating, positive and negative
5.	Non-horizontality	Cumulative, positive
6.	Sag	Cumulative, positive
7.	Variation of temperature	Cumulative, positive, negative
8.	Variation of pull	Compensating, positive

Permissible errors in chaining for measurements made with

1.	Invar tape, spring balances, thermometers	1 in 10,000
2.	Steel tape, plumb bob	1 in 2000
3.	Tested chain, plump bob	1 in 1000
4.	Chain under average conditions	1 in 500

Compass survey

It is a survey conducted with the help of prismatic or surveyors' compass.

Unit of measurement is an angle. It is the difference in directions of two intersecting lines.

Sexagesimal system

In this system

1 circumference = 360° (degrees)

1 degree = 60′ (minutes)

1 minute = 60″ (seconds)

Bearing

It is the horizontal angle which a line makes with some reference direction, known as meridian. The reference direction may be any of the following three—

(i) True meridian

(ii) Magnetic meridian

(iii) Arbitrary meridian

Bearing is of the following types—

(i) Whole circle bearing

(ii) Quadrantal bearing

(iii) Reduced bearing

(iv) Fore bearing

(v) Back bearing

Back bearing = Fore bearing ± 180°

Magnetic compass

Three types of compasses are commonly used which are—

(i) Prismatic compass

(ii) Surveyor's compass

(iii) Transit compass

Plane table survey

It is a method of surveying in which field work and office work are done simultaneously on a plane table. Following instruments are used in plane tabling.

(i) Plane Table

(ii) Alidade

(iii) Plumbing fork and plumb bob

(iv) Spirit level

(v) Compass

(vi) Rain proof cover for the plane table

(vii) Chain or tap

(viii) Ranging rods

(ix) Drawing sheets

(x) Drawing equipment

Methods of Plane Tabling

(i) **Radiation**—This method is used when distance are small.

(ii) **Intersection**—It is used when the distance between the point and the instrument station is either too large or can not be measured accurately due to field conditions.

(iii) Traversing

(iv) Re-section

The following are the four methods of orientations—

(i) Orientation by compass

(ii) Orientation by back sighting

(iii) Orientation by three point problem

(iv) Orientation by two point problem

Levelling

It is an important branch of surveying.

Principle of levelling

Essentially a level consists of the following five parts—

(i) A telescope to provide a line of sight

(ii) A level tube to make the line of sight horizental

(iii) Levelling head to bring the bubble in the centre of tube

(iv) A cross bubble tube to provide a horizontal plane

(v) A tripod to support the instrument

Types of levels

Following are commonly used levels—

(i) Dumpy level

(ii) Wye level

(iii) Cooke's reversible level

(iv) Cushing's level

(v) Modern Titing level of Indian office Pattern

Levelling Terminology

(i) Level surface

(ii) Level line

(iii) Horizontal surface

(iv) Horizontal line

(v) Vertical surface

(vi) Vertical line

(vii) Datum

(viii) Reduced level

(ix) Line of collimation
(x) Back sight
(xi) Fore sight
(xii) Intermediate sight
(xiii) Height of instrument
(xiv) Changing point
(xv) Station
(xvi) Bench mark
(xvii) Parallax
(xviii) Temporary adjustment of a level
(xix) Permanent adjustment of a level
(xx) Level book
(xxi) Reduced levels
(xxii) Reciprocal levelling

Contouring

A contour is an imaginary line on the ground joining the points of equal elevation.

Characteristics of Contours

(i) Two contours of different elevations do not cross each other except in the case of an over hanging cliff.

(ii) Contours of different elevations do not unite to form one contour except in the case of a vertical cliff.

(iii) Contours drawn closer depict a steep and if drawn far apart it represent a gentle slope.

(iv) Contours equally spaced depict a uniform slope.

(v) A contour at any point is perpendicular to the line of the steepest slope at the point.

(vi) A contour must close itself or go out of the limit of the map.

(vii) A set of ring contours with higher values inside, depict a hill whereas a set of ring contours with lower values inside, depict a pond or a depression.

(viii) When contours cross a ridge or V-shaped valley these form sharp V-shaped across them.

(ix) The same contour must appear an both the sides of a ridge or a valley.

Uses of Contour Maps

(1) To study the general character of the tract of the country without visiting the ground.

(2) To decide the most economical and suitable sites for engineering works such as canal, sewer, reservoir, road, railway etc.

(3) To determine the catchment area of the drainage basic and hence the capacity of the proposed reservoir.

(4) To compute the earth work required for filling or cutting along the linear alignment of projects such as canals, roads etc.

(5) To ascertain the intervisibility of the points.

(6) To trace a contour gradient for road alignments.

(7) To draw longitudinal sections and cross-sections to ascertain the nature of the ground.

(8) To calculate water capacities of reservoirs.

Minor instruments

The main minor instruments are as follows :

(i) Hand level
(ii) Clinometer
(iii) Abney's level
(iv) Tangent clinometer
(v) Ceylon chat tracer
(vi) Box sextant
(vii) Pantagraph
(viii) Planimeter

Tacheometric survey

Tacheometry is a branch of surveying in which the horizontal and vertical distances are determined by angular observations with a tacheometer. The chaining operations are altogether eliminated. Tacheometry is not as accurate as in chaining.

The main instruments used in tacheometry is a tacheometer. Tacheometric survey may be carried but by stadia hair system or tangential system.

Stadia hair system

This system may further be divided into two different methods—

(i) Fixed hair method

(ii) Movable hair method

There are four cases of tacheometric calculations. They are—

(i) Staff held vertical

(ii) Inclined sights staff held vertical

(iii) Inclined sights upwards, staff held normal

(iv) Inclined sights downwards, staff held normal

Tangential method of tacheometry

(i) Both angles being elevations

(ii) Both angles being depressions

(iii) One angle elevation and the other depression

Curves

Curves are regular bends provided in the lines of communication like roads, railways and canals to bring about the gradual change of direction. They are also used in the vertical plane at all changes of grade to avoid the abrupt change of grade at the apex. Curves provided in the horizontal plane to have the gradual change in direction are known as horizontal curves whereas those provided in the vertical plane to obtain the gradual change in the grade are known as vertical curves.

Triangulation

It is the process of measuring the angles of a network of triangles formed by stations marked on the surface of the earth.

Types of triangulation

(i) Primary triangulation

(ii) Secondary triangulation

(iii) Tertiary triangulation

Different terms used in triangulation

(i) True value

(ii) Observed value

(iii) True errors

(iv) Most probable error

(v) Residual error

(vi) Weight of the observations

(vii) Laws of weights

Astronomy

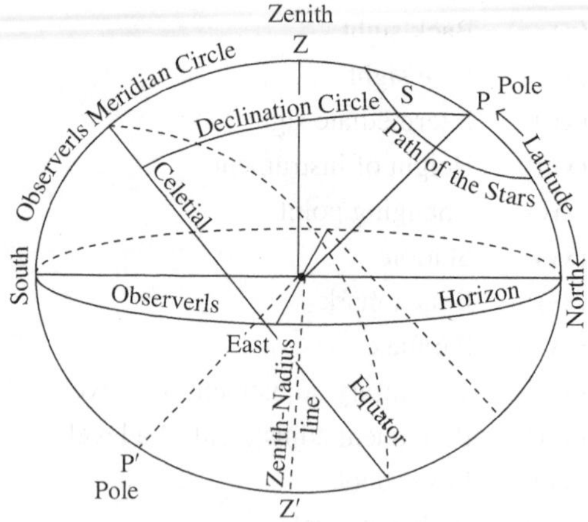

(i) **The celestial sphere**—The imaginary sphere on which heavenly bodies, *i.e.*, stars, sun, moon etc. appear to lie is known as celestial sphere.

(ii) **The Zenith**—The point on the celestial sphere above the observer's station is known as the Zenith.

(iii) **The Nadir**—The point on the celestial sphere exactly below the observer's station is known as the Nadir.

(iv) **The Zenith-Nadir line**—The Zenith, the observer's station the centre of the earth and the Nadir all lie on a line which is known as Zenith Nadir line.

(v) **The celestial horizon**—The great circle of the celestial sphere obtained by a plane passing through the centre of the earth and perpendicular to the Zenith-Nadir line is known as the celestial horizon.

(vi) **The visible horizon**—The small circle of the earth which is obtained by visual rays passing through the point of observation is known as visible horizon.

(vii) **The sensible horizon**—The small circle which is obtained by a plane-passing through the observer's station and tangential to the earth's surface and perpendicular to the Zenith-Nadir line at the point of observation is known as the sensible horizon.

(viii) **The terrestrial equator**—The great circle of the earth the plane of which is perpendicular to its axis of rotation is known as the terrestrial equator.

(ix) **The terrestrial poles**—The points at which earth's axis of rotation meets the earth's surface, are known as the terrestrial poles.

(x) **The celestial poles**—The points at which earth's axis of rotation on prolongation on either side meets the surface of the celestial sphere are known as celestial poles.

(xi) **Vertical circles**—The great circles of the celestial sphere passing through the Zenith and Nadir are known as vertical circles.

(xii) **The observer's meridian**—The vertical circle which passes through Zenith and Nadir of the station of the observation as well as through the poles, is known as observer's meridian.

(xiii) **The prime vertical**—The vertical which is perpendicular to the observer's meridian and passes through the east and west points of the horizon is known as the prime vertical.

(xiv) **North and south points**—The projections of the elevated north and depressed south poles on the horizon are known as north and south points respectively.

(xv) **East and west points**—The points on which the prime vertical meets the horizon, are known as east and west points. These points may also be obtained by the intersection of the equator and horizon.

(xvi) **Ecliptic**—The great circle of the celestial sphere which the sun appears to describe with earth as centre in the course of one year, is known as ecliptic.

(xvii) **Altitude**—The angular distance of a heavenly body above the horizon, measured on the vertical circle passing through it, is called altitude.

(xviii) **The azimuth**—The angle between the observer's meridian and the vertical circle passing through the celestial body and the Zenith is known as azimuth.

(xix) **The celestial latitude**—The arc of a great circle perpendicular to the ecliptic, intercepted between the celestial body and the ecliptic is known as the celestial latitude of the body.

(xx) **The celestial longitude**—The arc of an ecliptic intercepted between the great circle passing through the point of Aries (y) and the great circle passing through the celestial body is called celestial longitude.

360° of longitude = 24 hours of time
15° of longitude = 1 hour
15′ of longitude = 1 minute
15″ of longitude = 1 second
Apparent time = Mean time + Equation of time

Hydrographic survey

Survey carried out to depict accurately fathoms and other topographical features of large water bodies like sea rivers, lakes, bays etc. is known as Hydrographic survey.

The requirements of hydrographic survey

(i) Close soundings
(ii) Position of least depth on shoals
(iii) Position of breakers
(iv) Position of light vessels and buoys
(v) High and low water line
(vi) Tide tables
(vii) Measurement of ideal sream and currents
(viii) Observation magnetic variations
(ix) Continuous record of meteorological conditions

EXERCISE

1. The polar axis of earth is shorten than equatorial axis by—
   (A) 4295 metre (B) 9542 metre
   (C) 42·95 kilometre (D) None of these
2. In case of geographical maps, the scale is usually—
   (A) 1 cm = 10 cm to 10 cm
   (B) 1 cm = 50 m to 100 m
   (C) 1 cm = 5 km to 160 km
   (D) 1 cm = more than 200 km
3. If the smallest division of a vernier is longer than the smallest division of its primary scale, the vernier is known as—
   (A) Direct vernier
   (B) Double vernier
   (C) Simple vernier
   (D) Retrograde vernier
4. Hydrographic survey deal with the mapping of—
   (A) Large water bodies
   (B) Rainfall data

(C) Wave movement
(D) None of the above

5. The curvature of the earth's surface is taken into account only if the extent of survey is more than—
(A) 80 sq. km (B) 500 sq. km
(C) 260 sq. km (D) 1500 sq. km

6. Principle of surveying followed to prevent accumulation of errors is—
(A) To work from whole to the part
(B) To work from part to whole
(C) Both (A) and (B)
(D) None of the above

7. In centesimal system circumference is divided into—
(A) 400 parts (B) 360 parts
(C) 180 parts (D) 240 parts

8. The survey, in which the curvature of earth is ignored, is called as—
(A) Plane survey (B) Geodetic survey
(C) Geological survey (D) Aerial survey

9. The representative fraction is $\frac{1}{100000}$ signifies a scale of—
(A) 1 cm = 100000 m
(B) 1 cm = 10 kilometre
(C) 1 cm = 1 kilometre
(D) None of these

10. The representative fraction of the scale 1 cm = 1 kilometre will be—
(A) $\frac{1}{10000}$ (B) $\frac{1}{100}$
(C) $\frac{1}{1000000}$ (D) $\frac{1}{10000000}$

11. The scale, on which three dimensions can be measured, is known as—
(A) Plain scale (B) Diagonal scale
(C) Chord scale (D) Vernier scale

12. As per Indian standard, the length of one link in 30 metre chain should be—
(A) 30 cm (B) 20 cm
(C) 40 cm (D) 10 cm

13. A discrepancy is the difference between—
(A) True value and error
(B) Measured value and actual value
(C) Two measured values of the same quantity
(D) None of the above

14. Off sets are—
(A) Chain lines out of alignment
(B) Small measurements from chain line
(C) Measurements taken in chain surveying
(D) None of the above

15. When a chain is used at a temperature more than the temperature it was calibrated the error in measured length will be—
(A) Cumulative
(B) Compensating
(C) Negative
(D) Observational error

16. The length of an engineer's chain should be—
(A) 100 ft. (B) 60 ft.
(C) 30 m (D) 20 m

17. The linen tape reinforced with brass or copper wires to prevent streching or twisting of fibres is called—
(A) Limen tape (B) Metallic tape
(C) Invar tape (D) Cloth tape

18. Generally while chaining the arrow which accompany with one chain must be—
(A) 2 Nos. (B) 5 Nos.
(C) 10 Nos. (D) 12 Nos.

19. Method used for chaining on sloping ground is—
(A) By stepping method
(B) By hypotenusal allowance method
(C) By clinometer method
(D) Both (A) and (B) above

20. The error due to bad ranging is—
(A) Compensation
(B) Cumulative positive
(C) Cumulative negative
(D) Both (B) and (C) above

21. Negative errors are caused in chain, when its length is—
    (A) More than the standard length
    (B) Less than the standard length
    (C) Equal to the standard length
    (D) Any of the above

22. An error under the same conditions always of the same size and sign is called as—
    (A) Uniform error
    (B) Systematic error
    (C) Cumulative error
    (D) Any of (B) and (C) above

23. If n observations are taken, the accidental errors are proportional to—
    (A) n (B) n^2
    (C) $\sqrt{n}$ (D) $\frac{1}{n}$

24. The scale on which three dimensions can be measured is known as—
    (A) Plain scale
    (B) Diagonal scale
    (C) Chord scale
    (D) Vernier scale

25. As per Indian Standard the total number of links in a 30 metre chain should be—
    (A) 100 (B) 150
    (C) 300 (D) 75

26. A scale which has a common representative fraction, but read in different measures is called a—
    (A) Plain scale
    (B) Diagonal scale
    (C) Shrunk scale
    (D) Comparative scale

27. The error in measured length due to sag of chain or tape is known as—
    (A) Positive error
    (B) Negative error
    (C) Compensating error
    (D) Instrumental error

28. When the length of a chain along a slope of θ is (l) the correction for slope required is—
    (A) $l\,(1 - \sin\theta)$ (B) $l\,(1 - \tan\theta)$
    (C) $l\,(1 - \cos\theta)$ (D) $l\,(1 - \cos^2\theta)$

29. A triangle is said to be well conditioned if none of its angles is less than—
    (A) 20° (B) 30°
    (C) 45° (D) 60°

30. The longest chainline passing through the centre of the area is known as—
    (A) Base line (B) Tie line
    (C) Check line (D) All the above

31. The tie line is run through the survey to—
    (A) Take off sets for detailed surveying
    (B) Take details of nearby objects
    (C) Check accuracy of work
    (D) None of the above

32. The permissible limits of errors in chaining for measurement on rough or hilly ground is—
    (A) 1 : 2000 (B) 1 : 1000
    (C) 1 : 500 (D) 1 : 250

33. The position of point can be fixed more accurately by—
    (A) Perpendicular offset
    (B) Optical square
    (C) Prism square
    (D) Oblique offsets

34. The base line is measured with—
    (A) Metallic tape (B) Steel tape
    (C) Invar tape (D) Chain

35. Invar tapes are made of an alloy of—
    (A) Nickel and steel
    (B) Copper and steel
    (C) Tin and steel
    (D) Aluminium and steel

36. Metallic tapes are made of—
    (A) Steel
    (B) Invar
    (C) Cloth and wires
    (D) Nickel

37. An offset more than 15m in length is known as—
    (A) Long offset
    (B) Short offset
    (C) Oblique offset
    (D) Right-angled offset

38. An offset rod is provided at its top with—
(A) Hook
(B) Flag
(C) Distance measuring arrangement
(D) None of these

39. An obstacle which obstructs both chaining and ranging may be—
(A) River (B) Lake
(C) Hillock (D) Building

40. An obstacle which obstructs chaining but not ranging may be a—
(A) Building (B) Lake
(C) Hillock (D) Electric pole

41. The length of a ranging rod should be—
(A) 1·5 to 2 m (B) 2 to 3 m
(C) 3 to 4m (D) 2·5 to 4·5 m

42. The principle of working of an optical square is based upon—
(A) Refraction
(B) Reflection
(C) Double refraction
(D) Double reflection

43. The angle between two plane mirrors of an optical square should be—
(A) 30° (B) 45°
(C) 60° (D) 90°

44. The principle of optics used in the construction of an optical square is—
(A) Reflection
(B) Double reflection
(C) Refraction
(D) Double refraction

45. Which of the following instruments is used for measurement of angles ?
(A) Geodimeter (B) Tellurometer
(C) Sextant (D) Telescope

46. The direction of a line relative to a given median is called—
(A) Angular direction (B) Bearing
(C) Deviation (D) Inclination

47. True meridian passes through—
(A) Actual line of control
(B) A fixed point
(C) True north and true south
(D) Equatorial line

48. A horizontal angle which a line makes with the true meridian through one of the extremities of the line is called—
(A) Bearing
(B) True bearing
(C) Imaginary bearing
(D) Magnetic bearing

49. Which of the following instruments can be used for the direct measurement of directions?
(A) Surveyor's compass
(B) Prismatic compass
(C) Any of the above
(D) None of the above

50. The magnetic meridian at a point can be fixed by—
(A) Bar magnets
(B) Electro-magnets
(C) Magnetic compass
(D) None of the above

51. Traversing is generally done by—
(A) Chain
(B) Compass or theodolite
(C) Plane table
(D) All the above are correct

52. The box of prismatic compass is made of—
(A) Steel (B) Brass
(C) Iron (D) Aluminium

53. If 'n' is the number of sides, then the total sum of interior angles of a closed traverse should be—
(A) $n + 2$ (B) $2n - 4$
(C) $2n + 4$ (D) $4n - 2$

54. If n is the number of sides then the total sum of exterior angles of a closed traverse should be—
(A) $n + 2$ (B) $2n - 4$
(C) $2n + 4$ (D) $4n - 2$

55. If the end stations of a line are free from local attraction, then the difference in between fore and back bearing of that line should be—
(A) 180° (B) 120°
(C) 360° (D) 90°

56. When the meridian of a plane changes with time, then this meridian will be—
(A) Arbitrary (B) Magnetic
(C) True (D) Grid

57. The imaginary line joining the points of zero declination on the surface of the earth is called as—
(A) Isogonic line
(B) Isoclinic line
(C) Agonic line
(D) Magnetic declination line

58. The daily variation in magnetic declination is known as—
(A) Diurnal variation
(B) Solar variation
(C) Secular variation
(D) Irregular variation

59. The extent of daily variation of declination depends upon—
(A) The locality
(B) Season of the year
(C) Time of the day
(D) All of the above

60. The angle between the true meridian and the magnetic meridian at the time of observations is known as—
(A) Dip
(B) Bearing angle
(C) Magnetic bearing
(D) Magnetic declination

61. Irregular variation in magnetic declination may be due to—
(A) Magnetic storm
(B) Earthquakes
(C) Solar influence
(D) Any of the above

62. Magnetic declination at a point—
(A) Does not change with time
(B) Never remains constant
(C) Varies with time
(D) (B) and (C) above

63. The variation in magnetic declination may be—
(A) Diurnal variation
(B) Annual variation
(C) Secular variation
(D) Any of the above

64. The needle of magnetic compass is generally supported on a—
(A) Ball bearing (B) Bush bearing
(C) Needle bearing (D) Jewel bearing

65. At magnetic poles the dip of magnetic needle is—
(A) 0° (B) 45°
(C) 90° (D) 120°

66. The least count of prismatic compass is—
(A) 1° (B) 30′
(C) 15′ (D) 20″

67. The total number of zeroes graduated on a surveyor's compass is—
(A) One (B) Two
(C) Three (D) Four

68. At the equator, the amount of dip is—
(A) 0° (B) 45°
(C) 60° (D) 90°

69. A correction for error due to refraction is—
(A) $0{\cdot}01\ d^2m$ (B) $0{\cdot}001\ d^2m$
(C) $0{\cdot}01122\ d^2m$ (D) $0{\cdot}078\ d^2m$

70. The first temporary adjustment of a prismatic compass is—
(A) Levelling
(B) Focussing of the prism
(C) Removing the parallax
(D) Centring

71. Which of the following could be used for fixing the arbitrary direction ?
(A) River
(B) Moving train
(C) Road
(D) Top of a chimney

72. Arbitrary bearing is measured by—
(A) Theodolite
(B) Magnetic compass
(C) Astronomical observations
(D) Any of the above

73. What is the whole circle bearing of a quadrant bearing N 15° 28′ W ?
(A) 195°28′ (B) 164°32′
(C) 344°32′ (D) 344°28′

74. Irregular variation in magnetic declination may be due to—
(A) Magnetic storm (B) Earthquakes
(C) Solar influence (D) Lunar influence

75. The variation in magnetic declination is termed as—
(A) Diurnal variation (B) Solar variation
(C) Lunar variation (D) Secular variation

76. Magnetic meridian and true meridian are—
(A) Always parallel
(B) Parallel at the equator
(C) Parallel near the poles
(D) Never parallel

77. A prismatic compass is used to determine—
(A) Dip at a place
(B) Declination of a line
(C) Whole circle bearing of a line
(D) Quadrangle bearing of a line

78. The graduations in a prismatic compass are—
(A) Inverted and run clockwise having 0° at north
(B) Inverted and run clockwise having 0° at south
(C) Erect
(D) None of the above

79. Quadrantal system confines the user to measure the bearing from—
(A) The north point clockwise
(B) The north point or south point whichever is nearer
(C) The north point anticlockwise
(D) None of the above

80. A bearing noted 45° NE represents—
(A) Quadrantal system
(B) Whole circle system
(C) Reduced bearing system
(D) None of the above

81. The foremost work to be executed while traversing with a compass is—
(A) Marking and referencing of stations
(B) Running survey lines
(C) Picking up of details
(D) Reconnaissance

82. The temporary adjustment of a prismatic compass is—
(A) Centring
(B) Adjustment of levels
(C) Adjustment of needle
(D) Adjustment of vane

83. In a plane table survey the method used for locating details is—
(A) Resection (B) Radiation
(C) Intersection (D) (A) or (B) above

84. In plane table survey, the method used for locating plane table station is—
(A) Traversing (B) Resection
(C) Radiation (D) (A) or (B) above

85. In a plane table survey the plotting of inaccessible points can be conveniently done by—
(A) Method of interpolation
(B) Method of radiation
(C) Method of traversing
(D) Method of intersection

86. The two point or three point problems are methods of—
(A) Orientation
(B) Resection
(C) Traversing
(D) (A) and (B) above

87. The accuracy of fix obtained by solving a two point problem is—
(A) Good (B) Fairly accurate
(C) Not reliable (D) None of these

88. Accurate centring of a plane table is done by—
(A) Spirit level (B) Plumbing fork
(C) Alidade (D) Ranging rod

89. The working edge of an alidade is known as—
(A) Ebonite edge (B) Fiducial edge
(C) Straight edge (D) Graduated edge

90. The first temporary adjustment to be done after setting up the plane table at any station will be—
(A) Orientation
(B) Centring and levelling
(C) Levelling
(D) Sighting

91. The second temporary adjustment to be done after setting up the plane table at any station will be—
    (A) Levelling
    (B) Orientation
    (C) Centring and levelling done simultaneously
    (D) Sighting

92. In plane tabling, the straight lines, angles and triangles are constructed by—
    (A) Protractor
    (B) Divider
    (C) Graphical method
    (D) Planimeters

93. In plane table survey the method used for locating the plane table station is—
    (A) Radiation
    (B) Intersection
    (C) Radiation or intersection
    (D) Traversing

94. When the plane table is not correctly oriented, the resectors will—
    (A) Meet at a point
    (B) Form a triangle
    (C) Either of the above
    (D) None of the above

95. Any arbitrarily assumed level line from which vertical distances are measured is known as—
    (A) Level line (B) Vertical line
    (C) Horizontal line (D) Datum line

96. A staff reading taken on a bench mark or change point is known as—
    (A) Back sight
    (B) Fore sight
    (C) Intermediate sight
    (D) None of the above

97. The first observation taken on a turning point is—
    (A) Fore sight
    (B) Back sight
    (C) Both are correct
    (D) None of the above

98. The correction to be applied for eliminating the effect of earth's curvature is—
    (A) Additive (B) Subtractive
    (C) Compensating (D) Cumulative

99. In levelling the error due to refraction is—
    (A) Positive (B) Negative
    (C) Compensating (D) Progressive

ANSWERS

1. (B)	2. (C)	3. (D)	4. (A)	5. (C)
6. (A)	7. (A)	8. (A)	9. (C)	10. (C)
11. (B)	12. (A)	13. (C)	14. (B)	15. (A)
16. (A)	17. (B)	18. (C)	19. (D)	20. (B)
21. (A)	22. (D)	23. (C)	24. (B)	25. (B)
26. (D)	27. (A)	28. (C)	29. (B)	30. (A)
31. (B)	32. (A)	33. (A)	34. (D)	35. (A)
36. (C)	37. (A)	38. (A)	39. (D)	40. (B)
41. (B)	42. (D)	43. (B)	44. (B)	45. (C)
46. (B)	47. (C)	48. (B)	49. (C)	50. (C)
51. (D)	52. (B)	53. (B)	54. (C)	55. (A)
56. (B)	57. (C)	58. (A)	59. (D)	60. (D)
61. (D)	62. (D)	63. (D)	64. (D)	65. (C)
66. (C)	67. (B)	68. (A)	69. (C)	70. (D)
71. (D)	72. (A)	73. (C)	74. (A)	75. (A)
76. (B)	77. (C)	78. (A)	79. (B)	80. (A)
81. (D)	82. (A)	83. (D)	84. (D)	85. (D)
86. (D)	87. (C)	88. (B)	89. (B)	90. (B)
91. (C)	92. (C)	93. (C)	94. (B)	95. (D)
96. (A)	97. (A)	98. (B)	99. (C)	

Chapter-5

BUILDING MATERIALS AND CONSTRUCTION

The various types of materials used in building engineering structures are stones, soils, bricks, lime, cement, mortar, iron, paints, plastics, glass, asbestos, etc.

Rocks

Rocks are divided into three types—

(i) **Igneous rocks**—These are formed due to solidification of molten materials. These rocks are hard, tough, dense, strong and durable. These rocks may be (a) basic rocks containing less than 52% silica (b) acidic rocks containing more than 65% silica.

(ii) **Sedimentary rocks**—These are formed sediments through the agency of water, wind and glacier. Important of these rocks are sandstone, limestone, shale and laterite.

(iii) **Metamorphic rocks**—These are the secondary rocks formed by metamorphosis due to intense heat and pressure below the earth's crust.

Stones

Stones are the natural products of rocks and are directly used for building construction with little bit of work for sizing, finishing etc. Requirements of a good building stone are hardness, toughness, uniformity in textures and colour. Like of granite stone is taken to be 300 years and that of sandstone as 100 years. The life of stone is reduced by frost, wind, atmospheric impurities, temperature variation, vegetable growth etc. Principal building stones under use are granite, gneisses, Basalt, sandstone, limestone, marble, quartzite etc.

Bricks and tiles

These are blocks of tempered plastic moulded to suitable shapes. They are dried in sun and then burnt so as to make them strong, hard and durable. Bricks are classified into three parts—

Class (I) bricks—Characters :

(i) Uniform colour

(ii) Plane rectangular face

(iii) Parallel sides

(iv) Sharp straight right angled edges

(v) Compact

(vi) Uniform texture

Class (II) bricks—Characters :

(i) Slightly over burnt

(ii) Distorted round edge

(iii) Fine

(iv) Compact

(v) Uniform texture

Class (III) bricks—Characters :

(i) Slightly over burnt or under burnt

(ii) Defects of uniformity

Water absorption of bricks after 24 hours immersion in cold water should be upto

Class (I) bricks	—	20% by weight
Class (II) bricks	—	20% by weight
Class (III) bricks	—	25% by weight

The main constituents of earth for brick making are :

(i) Silica or sand

(ii) Alumina or clay

(iii) Lime

(iv) Magnesia

(v) Oxide of iron

Fire bricks

These bricks are generally made of fine clay or refractory clay and burnt at high temperature in special type of kiln. The composition of fire clay is—

(i) Silica	=	56 to 96%
(ii) Alumina	=	2 to 36%
(iii) Iron oxide	=	2 to 5%
(iv) Lime magnesia, potash and soda	=	8 to 20%

Tiles

Tiles are thin slabs used for convering roofs or flooring or for making drains. Tiles may be classified under three heads—

(i) Flooring or Paving tiles
(ii) Roofing tiles
(iii) Drain tiles

Manufacture of Bricks

There are three steps involved in it are—

(a) Preparation of earth which involves digging, weathering, blending and tempering.

(b) Moulding can be done by hand or machine

(c) Burning the moulded bricks in kiln

Lime

Lime is obtained by burning any form of lime stone over 900°C. Lime is used for plastering interior surfaces of walls, preparing concrete for foundation and flooring of ordinary buildings. It is used for washing the rooms, stablizing the soils and knotting of timber works before painting.

Quick lime

It is the calcined material, the major part of which is calcium oxide. It also contains relatively smaller amount of magnesium oxide and is capable of slating with water.

Fat lime

The lime which has high calcium oxide content and is dependent for setting and hardening solely on the absorption of carbon dioxide from the atmosphere.

Hydraulic lime

It is a lime containing small quantities of silica and alumina and/or iron oxide which are in chemical combinations with some of the calcium oxide content.

Hyderated lime

It is a dry powder resulting from the treatment of quick lime with water.

Cement

The product obtained by burning and crushing to powder either stone containing 1/5th or 2/5th clay and remaining carbonate of lime or an intimate mixture of well proportioned calcareous and argillaceous materials is called cement composition of Portland cement is as follows—

(a) Calcium oxide	=	60 to 65%
(b) Silica	=	20 to 25%
(c) Aluminium oxide	=	4 to 8%
(d) Ferrous oxide	=	2 to 4%
(e) Magnesium oxide	=	1 to 3%
(f) Sulphur Trioxide	=	0·5 to 3%

Initial setting time

It is the time for attaining that stage in the process of hardening after which any cracks that may appear do not reunite.

Final setting time

It is the time for attaining sufficient strength and hardness.

The initial setting time for cement should not be less than 30 minutes and the final setting time not more than 10 hours.

Different Kinds of Cement

(1) **Quick setting cement**—This type of cement has initial setting time of 5 minutes and final setting time of 30 minutes.

(2) **Rapid hardening cement**—This type of cement contains more of tri-calcium silicate and gains strength faster than the ordinary cement.

(3) **High alumina cement**—It is manufactured by fusing together a mixture of bauxite and limestone in correct proportions and at high temperatures.

(4) **Low heat cement**—It has high percentage of tricalcium aluminate and tricalcium silicate.

(5) **White cement**—It is manufactured from white chalk and clay free from iron oxide. It is white in colour.

Coarse Aggregates

Aggregates are the inert materials which when bound together by cement, will form concrete. Aggregates should comply with the requirements of IS 383. Aggregates used in India are the naturally occurring aggregates like sand, crushed rock and gravel.

Qualities of aggregates—

(i) Good sand should have coarse and angular gains of pure silica.

(ii) The gains of sand should be hard, strong and durable.

(iii) It should be free from silt, clay or any such salt that may hamper setting or attack the reinforcement.

(iv) It should not contain any organic matter.

(v) It should be well graded.

Asphalt

Asphalt is a natural or artificial mixture in which bitumen is associated with inert mineral matter.

Products of asphalt are—

(i) Asphalt cut book

(ii) Asphalt emulsion

(iii) Asphalt cement

(iv) Mastic asphalt

Paints

The fluid paste prepared by dissolving a base into a vehicle (carrier) along with a colouring pigment is known as paint.

Uses—

(i) It protects the different surfaces from weathering effects of the atmosphere and to keep them in good condition for a long period.

(ii) It prevents decay of wooden members by wet rot.

(iii) It provides colourful, pleasant and decorative appearance to the surfaces.

Types of paints—

(i) Oil paints
(ii) Enamel paints
(iii) Cement paints
(iv) Plastic paints
(v) Water paints
(vi) Emulsion paints
(vii) Bituminous paints
(viii) Aluminium paints
(ix) Distempers

Masonry

The construction work made by stone or brick is termed as masonry structure. It is divided into two classes—

(i) Stone masonry (ii) Brick masonry

(i) **Stone masonry**—When rock is quarried out and given rough shape at the quarries it is called stone masonry. It is divided into—

(a) Dry masonry

(b) Ashlar masonry

(c) Rubble masonry

(ii) **Brick masonry**—The construction done with bricks bonded together with the help of a mortar is termed as brick masonry. The bricks are throughly soaked in water for atleast one hour before use.

Backing—It is the inside of a wall.

Bat—A portion of a brick cut across the width is called bat.

Bed—It is the flat bottom side of the brick.

Bond—The method of arranging bricks so that the individual units are tied together is called bond.

Closer—A portion of a brick cut longitudinally with one long face uncut.

Course—It is the layer of bricks laid on the same bed.

Cow nose—The brick with two rounded edges used in quoin etc.

Facing—Exposed surface of a wall.

Hearting—Part of the wall between facing and backing.

Queen closer—It is placed next to the first brick in a header. This is half brick cut longitudinally.

King closer—It is used for splaying brick work.

Bevelled closer—It has one stretcher face closed.

Frogs—These are indentations on the bricks so as to form a key for holding the mortar.

Header—This denotes the end of a brick as seen in the wall face.

Quoin—A corner or the external angle of a wall.

Stretcher—The side of a brick as seen in elevation in a wall where the brick is laid flat.

Bed surface—The surface of a stone perpendicular to the line of pressure.

Corbel—A piece of stone projecting from a wall to support a structural member.

String course—This is a continuous course of masonry projecting from the face of the wall and is provided to throw off water.

Templates—These are pieces of stone placed under the end of a beam to distribute the load over greater area.

Throating—Grooves cut on the under surfaces of sills, copings, string courses etc. to prevent the water from trickling down the walls.

Joints in ashlar masonry—Since ashlar masonry is built up of stones carefully dressed, it has narrow joints.

Types of Bond	Salient place of use
1. Stretcher bond	Suitable for half brick position of wall
2. Header bond	Suitable for one brick wall and curved wall
3. English bond	Suitable for all types
4. Flemish bond	
(i) Single	Minimum wall thickness of one and half bricks
(ii) Double	Economical than English bond but not stronger.
5. Garden wall bond	
(i) English	Suitable for garden and boundary walls.
(ii) Flemish	Is also known as scotch or sussex bond
6. Facing bond	Suitable for thick walls with facing and backing
7. Dutch bond	Modified english bond with stronger corners.
8. Zigzag bond	Suitable for bricks floor paving.
9. Raking bond	
(i) Herring bone	For wall thickness of more than four bricks.
(ii) Diagonal	For wall thickness of two or four bricks.
10. English cross bond	Modified english bond with enhanced beauty and strength
11. Brick-on-edge bond	Suitable for partition wall.
12. Bonds at junction	Suitable for intersecting walls

Loads

A structure is always designed with a fair degree of accuracy to withstand various loads containing on it.

(i) **Dead load**—It is due to weight of all walls, partitions, floors, roofs.

(ii) **Live load**—All loads other than dead load are taken as live loads. Live loads on roofs are due to rain, wind, impact vibration etc.

Bearing Capacity of Foundation Material

Type of Rock/soil	Max. safe bearing capacity (in tonnes per sq. m.)	
Rocks		
(a) Igneous rocks	1000	
(b) Lime stone and hard stone	400	
(c) Schists and Slales	300	
(d) Clay shales	100	
Non Choesive Soils	**Dry**	**Submerged**
(a) Compact sand	40–60	20–30
(b) Loose well graded sand	20–40	10–20
Cohesive Soils		
(a) Stiff clay and sandy clay	20–40	
(b) Firm Clays	10–20	
(c) Soft clays and silts	5–N%	

Foundation—Foundation is that part of the building which supports the super and transmits its load to the sub-soil below it. Various types of foundations are—

(i) Spread foundation

(ii) Stepped or benching foundation

(iii) Grillage foundation

(iv) Pile foundation

(v) Pier (Well) foundation

(vi) Caisson's foundation

Damp Proof Course—(DPC) A layer provided to prevent entry of unwanted moisture inside the building either by seepage or by leakage is known as Damp Proof Course.

The dampness is injurious to the materials which come in its contact. The various effects of dampness are the following :

(i) Dampness breed bacteria of various diseases.

(ii) It causes decay of all the materials which come in contact, specially timber and steel.

(iii) It destroys furnishings in a building such as carpets, door frames, electrical fittings etc.

(iv) It causes disintegration of brick work tile, stone etc.

(v) It also creates health hazard to living beings.

Flooring

The structures which divide a building into stages or storeys to give the space to live at different levels are termed as floors.

Types of floors

(i) Timber floors

(ii) Cement tile concrete flooring

(iii) Cement concrete flooring

(iv) Mosaic floor

(v) Brick flooring

(vi) Flag stone flooring

Lintels—These are load carrying members usually made of R.C.C. or stone. Load carried by lintels depends on—

(i) Height of the wall above lintels

(ii) Position of openings

(iii) Position of concentrated loads

Roofs

To keep out rain, sun and wind and to preserve the interior from exposure, the horizontal surface used over most part of the building is called roof. There are two types of roofs—

(i) Flat roof

(ii) Pitched roof

Plastering

It is the process of converting rough surfaces of various structures by a plastic materials to form an even, smooth and durable surface.

Following are the types of plaster—

(1) Limeplaster

(2) Cement plaster

(3) Stucco plaster

(4) Water proof plaster

(5) Plaster on lath

Scaffolding

The temporary network errected to support a number of platforms at different levels for the convenience of workers is called scaffolding various types of scaffolding are as follows—

(1) Single scaffolding

(2) Double scaffolding

(3) Cantilever scaffolding

(4) Suspended scaffolding

(5) Trestle scaffolding

(6) Steel scaffolding

EXERCISE

1. Most weather resisting metamorphic rock is—
(A) Limestone (B) Slate
(C) Marble (D) Quartzite

2. Sandstone mostly contains—
(A) Quartz and lime
(B) Quartz, lime and silica
(C) Silica, alumina and lime
(D) None of these

3. Laterite is—
(A) Argillaceous rock
(B) Silicious rock
(C) Calcareous rock
(D) Magma

4. Sandstone is—
(A) Volcanic rock
(B) Igneous rock
(C) Metamorphic rock
(D) Sedimentary rock

5. Quartzite is—
(A) Silicious rock
(B) Argillaceous rock
(C) Metamorphic rock
(D) Calcareous rock

6. Basalt is classified as—
(A) Extrusive igneous rock
(B) Sedimentary rock
(C) Metamorphic rock
(D) Intrusive igneous rock

7. Chemically marble is a—
(A) Calcareous rock
(B) Silicious rock
(C) Argillaceous rock
(D) None of the above

8. Marble is—
(A) Igneous rock
(B) Metamorphic rock
(C) Stratified rock
(D) Sedimentary rock

9. Marble contains—
(A) Lime (B) Lime and silica
(C) Silica (D) None of these

10. Rocks formed by gradual deposition are—
(A) Sedimentary rocks
(B) Metamorphic rocks
(C) Igneous rocks
(D) None of these

11. The colour of a mineral in its powdered form is known as—
(A) Streak (B) Lusture
(C) Texture (D) Structure

12. Which of the following is a mineral ?
(A) Quartzite (B) Laterite
(C) Granite (D) Calcite

13. Slate is formed from the metamorphic action on—
(A) Shale (B) Limestone
(C) Sandstone (D) Granite

14. Which of the following is a rock ?
(A) Gypsum (B) Augite
(C) Calcite (D) Mica

15. Rocks having silica as main constituent are called—
(A) Calcareous rocks
(B) Argillaceous rocks
(C) Silicious rocks
(D) None of these

16. The rock having calcium carbonate as main mineral constituent, is known as—
    (A) Calcareous rock
    (B) Argillaceous rock
    (C) Silicious rock
    (D) Sandy rock
17. Kaoline is a—
    (A) Calcareous rock
    (B) Argillaceous rock
    (C) Silicious rock
    (D) Sandy rock
18. Trap is a—
    (A) Metamorphic rock
    (B) Igneous rock
    (C) Granite rock
    (D) Sedimentary rock
19. The minimum hardness number for marble is about—
    (A) 3 (B) 8
    (C) 5 (D) 10
20. Granite contains primarily—
    (A) Quartz
    (B) Quartz and feldspar
    (C) Mica
    (D) Quartz and mica
21. Common sand is a variety of—
    (A) Calcite (B) Quartz
    (C) Feldspar (D) Chlorite
22. A stone is rejected if the water absorption is more than—
    (A) 15% (B) 20%
    (C) 25% (D) 30%
23. For carving ornamental and architectural beauty the stone should be—
    (A) Soft (B) Hard
    (C) Heavy (D) Light
24. Stone generally used for railway ballast is—
    (A) Sandstone
    (B) Dolomite
    (C) Marble
    (D) Basalt or trap basalt
25. Attrition test on stone is done to find out—
    (A) Compressive strength
    (B) Hardness
    (C) Rate of water
    (D) Toughness
26. A good building stone should be—
    (A) Strong
    (B) Good in appearance and colour
    (C) Hard and tough
    (D) All the above
27. The stone that exhibits highest compressive strength is—
    (A) Slate (B) Gneisses
    (C) Granite (D) Limestone
28. The appearance of a mineral in reflected light is known as—
    (A) Streak (B) Lusture
    (C) Texture (D) Structure
29. The coarseness of the grains of a mineral is known as—
    (A) Fracture (B) Texture
    (C) Structure (D) Lusture
30. The strength of a building stone should not be less than—
    (A) 1000 kg/cm^2 (B) 500 km/cm^2
    (C) 1500 kg/cm^2 (D) 300 km/cm^2
31. The porosity of a good building stone should not be less than—
    (A) 3% (B) 5%
    (C) 10% (D) 20%
32. The tool used for quarrying of stones is—
    (A) Dipper
    (B) Spalling hammer
    (C) Jumper
    (D) Priming needle
33. The compressive strength of stone is found out by—
    (A) Attrition test (B) Crushing test
    (C) Hardness test (D) Impact test
34. The building stone used for bridge piers and columns is—
    (A) Granite (B) Limestone
    (C) Marble (D) Sandstone

35. The soil good for making bricks is—
(A) Black cotton (B) Alluvial
(C) Silty (D) Sand and silt

36. The impurity not desirable in the soil used for brick formation is—
(A) Alkali
(B) Kankar
(C) Both (A) and (B)
(D) None of these

37. Red colour is imparced to bricks due to—
(A) Iron oxide (B) Lime
(C) Silica (D) Magnesia

38. The weight of common bricks varies from—
(A) 1200 to 1400 kg/m^3
(B) 1400 to 1600 kg/cm^3
(C) 1600 to 1920 kg/m^3
(D) 2000 kg/m^3

39. The number of bricks required per cubic metre of brick masonry is—
(A) 400 (B) 450
(C) 500 (D) 550

40. Percentage of silica in a good brick earth lies between—
(A) 10–20% (B) 20–40%
(C) 50–60% (D) None of these

41. The red colour of the brick is due to—
(A) Iron oxide (B) Silica
(C) Magnesia (D) Alumina

42. The percentage of alumina in good brick earth lies between—
(A) 10–20% (B) 20–30%
(C) 30–40% (D) 40–50%

43. Excess of silica in brick earth causes—
(A) Loss of cohesion
(B) Impermeability
(C) Cracking and warping on drying
(D) Brittleness

44. The main function of alumina in brick earth is to—
(A) Impart plasticity
(B) Increase durability
(C) Prevent shrinkage
(D) Increase impermeability

45. The process of mixing clay, water and other ingredients to make brick is called—
(A) Moulding (B) Tempering
(C) Pugging (D) Blending

46. Lime is mixed with brick earth to—
(A) Impart plasticity
(B) Increase durability
(C) Prevent shrinkage
(D) Increase impermeability

47. The crushing strength of I class brick should not be less than—
(A) 35 kg/cm^2 (B) 70 km/cm^2
(C) 105 kg/cm^2 (D) 140 kg/cm^2

48. The minimum compressive strength of II class brick should be—
(A) 35 kg/cm^2 (B) 70 kg/cm^2
(C) 105 kg/cm^2 (D) 140 kg/cm^2

49. The temperature at which bricks are burnt varies from—
(A) 400° to 700° C
(B) 700° to 1000° C
(C) 1000° to 1200° C
(D) 1200° to 1500° C

50. For one cubic metre of brick masonry the number of bricks required is—
(A) 300 (B) 500
(C) 1000 (D) 2000

51. The bricks used for lining of furnaces are—
(A) Under burnt bricks
(B) Over burnt bricks
(C) Refractory bricks
(D) None of these

52. Fire bricks should be laid in—
(A) Lime mortar
(B) Fire cement mortar
(C) Fire clay mortar
(D) All the above

53. In brick masonry the frog of the brick is generally kept on—
(A) Top face (B) Bottom face
(C) Exposed face (D) Interior face

54. Bullnose bricks are not used in—
(A) Walls
(B) Rounding off sharp corners
(C) Arches
(D) Pillars

55. Hollow bricks are used for—
(A) Ornamental designs
(B) Thermal insulation
(C) Reduction in cost
(D) Earthquake resistant buildings

56. Fire bricks are used—
(A) To reflect heat
(B) To increase heat flow
(C) To decrease heat flow
(D) None of these

57. The board used for providing frog in bricks is known as—
(A) Pallel board (B) Lamin board
(C) Fibre board (D) Stock board

58. The approximate weight of a brick is—
(A) 2·0 kg (B) 3·2 kg
(C) 4·0 kg (D) 5·0 kg

59. Terracota is used in buildings for—
(A) Ornamental work (B) Insulation
(C) Storage (D) None of these

60. Swelling of bricks is known as—
(A) Bloating (B) Lamination
(C) Chuffs (D) Efflorescence

61. The maximum percentage of fine sand content in soil for making good tiles is about—
(A) 20 (B) 30
(C) 40 (D) 50

62. The maximum percentage of clay content for making good tiles is about—
(A) 20 (B) 30
(C) 40 (D) 50

63. The maximum percentage of silt content for making good tiles is about—
(A) 20 (B) 30
(C) 40 (D) 50

64. The process of burning limestone to redness in contact with air is called—
(A) Calcination (B) Slaking
(C) Hydration (D) None of these

65. Quick lime is—
(A) Calcium oxide
(B) Calcium hydroxide
(C) Calcium carbonate
(D) None of these

66. Quick lime—
(A) Generates heat on water addition
(B) Reacts with carbon dioxide
(C) When mixed with water forms slaked lime
(D) All the above

67. Plaster of Paris is obtained from the calcination of—
(A) Bauxite (B) Limestone
(C) Dolomite (D) Gypsum

68. Very strong lime mortar is obtained from—
(A) Quick lime
(B) Semi-hydraulic lime
(C) Hydrated lime
(D) Hydraulic lime

69. Fat lime can be used in—
(A) Distempers (B) Lime mortars
(C) Lime terracing (D) None of these

70. Lime putty is a—
(A) Paste of slaked lime
(B) Paste of lime and fine sand
(C) Paste of lump lime
(D) None of these

71. Fat lime is also called—
(A) Quick lime (B) Hydraulic lime
(C) Slaked lime (D) White lime

72. Quick lime is obtained from the burning of—
(A) Kankar (B) Lime stone
(C) Slaked lime (D) All the above

73. As compared to fat lime, the hydraulic lime sets—
(A) Faster (B) Slowly
(C) Same rate (D) Very slowly

74. A thin purable suspension of slaked lime in water is known as—
(A) Lime water (B) Milk of lime
(C) Lime putty (D) Hydrated lime

75. Plaster of Paris is obtained from the calcination of—
(A) Kankar (B) Lime stone
(C) Bauxite (D) Gypsum

76. The initial setting time for class 'A' lime is—
(A) 15 minutes (B) 30 minutes
(C) 120 minutes (D) 600 minutes

77. The advantage of adding pozzolana to lime is to—
(A) Reduce shrinkage
(B) Increase resistance to cracking
(C) Impact greater strength
(D) All the above

78. Volume of 1 bag of cement weighing 50 kg is—
(A) 3·4 cu m (B) 0·034 cu m
(C) 1·05 cu m (D) 0·020 cu m

79. While preparing lime mortar, the general ratio adopted in between lime and Surkhi or sand is—
(A) 1 : 1 (B) 1 : 2
(C) 1 : 4 (D) 1 : 6

80. Compared to mild steel, cast iron has—
(A) High compressive strength
(B) Low tensile strength
(C) (A) and (B)
(D) None of the above

81. Stainless steel resists corrosion due to—
(A) Carbon (B) Manganese
(C) Chromium (D) Sulphur

82. The percentage of carbon in cast iron is—
(A) 0·7% (B) 50%
(C) 1·2 to 1·7% (D) 2 to 4·5%

83. The percentage of carbon in steel is—
(A) 0·1 to 0·2% (B) 0·15 to 1·5%
(C) 1 to 3% (D) 2 to 4·5%

84. The maximum percentage of carbon in wrought iron is—
(A) 0·15% (B) 0·9%
(C) 1·5% (D) 3%

85. A short log 1·25 m or less in length is called—
(A) Pale (B) Bale
(C) Bolt (D) Post

86. A thin plank less than 5 cm thickness and more than 10 cm width is called a—
(A) Batten (B) Plank
(C) Board (D) Scantling

87. The decay of timber caused by alternate wetting and drying is called—
(A) Dry rot (B) Wet rot
(C) Foxiness (D) Doatiness

88. Plywood is specified by its—
(A) Weight
(B) Volume
(C) Thickness
(D) None of the above

89. A solvent commonly used for oil paints is—
(A) Tar (B) Turpentine
(C) Petrol (D) Aqua regia

90. Rusting can be prevented by applying coating of—
(A) Red lead (B) White lead
(C) Zinc oxide (D) Bitumen

91. Turpentine oil is used in paints as—
(A) Base (B) Thinner
(C) Vehicle (D) Drier

92. Linseed oil is used in paint as—
(A) Base (B) Vehicle
(C) Drier (D) None of these

93. The main constituent of varnish is—
(A) Turpentine oil (B) Petrol
(C) Resin (D) Solvent

94. Duco paint is a kind of—
(A) Bituminous paint
(B) Cellulose paint
(C) Emulsion paint
(D) Aluminium paint

95. French polish is—
(A) Oil paint
(B) Distemper
(C) Emulsion paint
(D) Spirit varnish

96. The base material for distemper is—
(A) Lime (B) Chalk
(C) Snowcem (D) Cement

97. Snowcem is—
(A) A distemper
(B) A synthetic enamel paint
(C) A primer
(D) A water proof cement paint

98. In paints, pigment is added for—
(A) Shining surface
(B) Colour
(C) Smoothening the surface
(D) Easy drying

99. The specific gravity of bitument is—
(A) 1·80 (B) 1·0924
(C) 1·00 (D) 0·80

100. On heating, bitumen becomes—
(A) Hard (B) Soft
(C) No change (D) None of these

101. Screws for wood work as specified by—
(A) Length (B) Diameter
(C) Weight (D) Gauge

102. In king post the number of vertical posts is—
(A) One (B) Two
(C) 3 or more (D) An even number

103. A rolling shutter is often used for—
(A) Drawing rooms (B) Cinema halls
(C) Beauty parlours (D) Shops

104. Different types of pitched roofs are—
(A) Mansard type (B) Gable type
(C) Both (A) and (B) (D) Thatch type

105. The roof having slope in all four direction is called—
(A) Hip-pitch roof (B) Shed roof
(C) Gambrel roof (D) North light roof

106. Apex line of a sloping roof is called—
(A) Rise (B) Hip
(C) Ridge (D) Verge

107. Lower edge of the inclined roof surface is called—
(A) Pitch (B) Eves
(C) Cleat (D) Gable

108. The shell roof may be designed as—
(A) Thin shell only
(B) Thick shell only
(C) Both thin shell and thick shell
(D) None of these

109. The head room in a staircase should not be less than—
(A) 3·5 m (B) 3 m
(C) 2·10 m (D) 2 m

110. The joints provided in wooden floor are—
(A) Tongue and groove
(B) Butt
(C) Dove tail
(D) Tee joint

111. Brick floors are used in—
(A) Workshops
(B) Railway platforms
(C) Godowns
(D) Verandahs

112. In multistoreyed government buildings, the floor height is usually restricted to—
(A) 2 m (B) 2·4 m
(C) 3·4 m (D) 4·2 m

113. Broken bricks or stone chips are also called—
(A) Brokomite (B) Scrap
(C) Spall (D) Wastage

114. In residential buildings the height of bathroom should not be less than—
(A) 2·0 m (B) 2·4 m
(C) 3·0 m (D) 3·6 m

115. The weight of river sand per cubic metre is taken as—
(A) 1000 kg (B) 1200–1400 kg
(C) 1800–1900 kg (D) 2200–2400 kg

116. The lime-sand mortar normally used in brick construction has a ratio of—
(A) 1 : 8 (B) 1 : 6
(C) 1 : 4 (D) 1 : 2

117. The ratio of cement-sand mortar preferred in load bearing walls is—
(A) 1 : 2 (B) 1 : 6
(C) 1 : 8 (D) 1 : 12

118. The number of steps in a flight should not be less than—
(A) 12 (B) 3
(C) 2 (D) 5

119. The number of steps in an ordinary flight should not be more than—
(A) 12 (B) 10
(C) 15 (D) 20

120. The width of landing should not be less than—
(A) 1·0 m (B) 1·2 m
(C) 0·9 m (D) Stair width

121. The sum of tread and rise should lie between—
(A) 35–40 cm (B) 50–55 cm
(C) 40–45 cm (D) 55–60 cm

122. Dog legged stairs are—
(A) Straight stair
(B) Quarter turn stair
(C) Half turn stair
(D) Three fourth stair

123. Half turn stairs change their direction through—
(A) 180° (B) 160°
(C) 90° (D) 45°

124. In public buildings the maximum riser is limited to—
(A) 10 cm (B) 12 cm
(C) 15 cm (D) 18 cm

125. A step approximately triangular in shape is known as—
(A) Angular steps (B) Radial steps
(C) Winders (D) Spandril

ANSWERS

1. (D)	2. (A)	3. (A)	4. (D)	5. (A)
6. (A)	7. (A)	8. (B)	9. (B)	10. (A)
11. (A)	12. (D)	13. (A)	14. (A)	15. (C)
16. (A)	17. (B)	18. (B)	19. (A)	20. (B)
21. (B)	22. (A)	23. (B)	24. (D)	25. (C)
26. (D)	27. (B)	28. (B)	29. (B)	30. (A)
31. (B)	32. (C)	33. (B)	34. (A)	35. (A)
36. (C)	37. (A)	38. (C)	39. (C)	40. (C)
41. (A)	42. (B)	43. (A)	44. (A)	45. (B)
46. (C)	47. (C)	48. (B)	49. (B)	50. (B)
51. (C)	52. (C)	53. (A)	54. (C)	55. (B)
56. (C)	57. (D)	58. (B)	59. (A)	60. (A)
61. (C)	62. (B)	63. (B)	64. (A)	65. (A)
66. (D)	67. (C)	68. (D)	69. (A)	70. (A)
71. (D)	72. (B)	73. (D)	74. (B)	75. (D)
76. (C)	77. (D)	78. (B)	79. (B)	80. (C)
81. (C)	82. (D)	83. (B)	84. (A)	85. (C)
86. (C)	87. (B)	88. (C)	89. (B)	90. (A)
91. (B)	92. (B)	93. (C)	94. (B)	95. (D)
96. (B)	97. (D)	98. (B)	99. (B)	100. (B)
101. (A)	102. (A)	103. (D)	104. (C)	105. (A)
106. (C)	107. (B)	108. (C)	109. (C)	110. (A)
111. (C)	112. (C)	113. (C)	114. (B)	115. (C)
116. (D)	117. (B)	118. (B)	119. (A)	120. (D)
121. (C)	122. (C)	123. (A)	124. (C)	125. (D)

●●●

Chapter-6

SOIL MECHANICS

Soil Mechanics

It is the branch of physical science which deals with the properties, nature and performance of soil.

Soil

Soil is a non-homogeneous, porous earthen mixture of inorganic material (consisting of soft clays, loams, silts and gravel or hard stiff clay in various proportions) formed by disintegration and decomposition of rocks.

On the geological basis the soils are classified into two main groups.

(i) **Residual soil**—It is found at the place of disintegration of the parent rock. Igneous rocks such as granite or basalt and sedimentary rocks such as sandstone, shale or limestone are the parent rock material for residual soils.

(ii) **Transported soil**—It is carried away from its place of disintegration to some other place by the transporting agencies like force of gravity, water, wind, ice etc.

On the basis of transportation and deposition the transported soils may be classified as—

(i) **Alluvial soil**—It is transported from its place of disintegration by the flowing water and deposited along the stream.

(ii) **Marine soil**—It is transported by water and deposited at the mouth of the stream.

(iii) **Lacustrine soil**—It is fine grained soil transported by water and deposited at the bottom of lakes.

(iv) **Loess or aeolian soil**—It is wind blown yellowish light brown dust consisting of silt or fine sand particles.

(v) **Dune sand**—It is a low hill of sand on the sandy shores of sea, rivers and lakes which develops due to loose sand being exposed to wind.

(vi) **Colluvial soil**—It is the soil collected at the base of a steep cliff because of the effect of the force of gravity on the broken rock fragments.

(vii) **Glacial drift or stratified drift**—It is the name given to the material picked up, mixed transported and redeposited by glacial melt water.

Main categories of soils are—

(i) Non-cohesive coarse-grained soils

(ii) Cohesive fine-grained soil

(iii) Organic soils

In 1911 Atterberg, a Swedish scientist, defined the boundaries of the four states in terms of the following limits—

Liquid limit is the boundary between the liquid and plastic states.

Plastic limit is the boundary between the plastic and semi-solid states.

Shrinkage limit is the boundary between the semi-solid and solid states.

Liquid limit is the minimum moisture content at which the soil is still in liquid state, but has started exhibiting a small shear strength against flowing.

Plastic limit is the minimum moisture content at which the soil can be rolled into 3 mm threads without showing any sign of cracks.

Shrinkage limit is the maximum moisture content at which a reduction in water content will not cause a decrease in volume of the soil.

Plasticity index = Liquid – Plastic limit

Permeability

Permeability is defined as the property of a porous material which permits the passage or seepage of water through its inter-connecting voids.

Permeability of soil depends upon—

(i) Grain size of soil
(ii) Void ratio
(iii) Structural arrangement of soil particles
(iv) Stratification
(v) Entrapped air and foreign matter
(vi) Properties of percolating water
(vii) Properties of the pore water and degree of saturation
(viii) Absorbed water in clayes soils

The flow of water through soils may be—

(i) **Laminar (linear) flow**—In which all water particles move in parallel paths without crossing the paths of other particles.

(ii) **Turbulent (eddy) flow**—In which water particles move along zig-zag paths and cross the path of other properties.

The velocities of flow of soil water are classified on—

(i) **Discharge velocity**—The rate of discharge of soil water through unit area of cross-section of the soil sample (both the areas of solids and voids) is known as discharge velocity.

(ii) **Seepage velocity**—The rate of discharge of percolating water per unit cross-sectional area of voids only perpendicular to the direction of flow, is known as seepage velocity.

Determination of coefficient of permeability of soil samples—

The permeability of soil samples may be determined by

(i) Falling head permeameter
(ii) Constant head permeameter

Seepage—The pressure which is exerted by water on the soil through which it percolates is called seepage pressure. Depending upon the direction of flow. Seepage may increase or decrease the vertical effective pressure of a soil.

Flow net—Graphical presentation of water flow through the soil voids is called flow net. Its knowledge helps in knowing quicksand condition, uplift water pressure and seepage amount etc.

Fundamental characteristics of a flow net—

1. Flow lines and equipotential lines in a flow net intersect at right angles.
2. All fields formed by equipotential lines and flow lines are approximate squares.
3. Same quantity of flow occurs through each flow channel.
4. Same potential drop occurs between any two successive equipotential lines.
5. The smaller the dimension of a field, greater is the hydraulic gradient and also velocity of flow.
6. In homogeneous soils, the shape of curves is either elliptical or parabolic.

Compaction—The process of making the soil dense is called compaction. Compaction may be accomplished by compressing the soil elastically reduction of air space and removing the pore water.

Different factors affecting compaction are—

(i) Type of soil
(ii) Moisture content
(iii) Means of compaction
(iv) Compaction effort

Consolidation—Consolidation is the gradual reduction in volume of a soil mass resulting from an increase in and continued application of compressive stress and is due to expulsion of water from the pores.

Types of consolidation

(i) Initial consolidation
(ii) Secondary consolidation
(iii) Primary consolidation

Terzaghi's one dimensional consolidation theory

The one dimensional consolidation theory proposed by Karl Terzaghi in 1923, is based on the following assumptions—

1. The soil is homogeneous.
2. The soil is 100% saturated.
3. The drainage and consolidation both are one dimensional.
4. The compressibility of soil grains and water is practically nil.
5. The soil is laterally confined.
6. The Darcy's law is void.
7. The coefficient of permeability is constant during the process of consolidation.
8. The time lag of consolidation is entirely due to low permeability of the soil.
9. The change in the effective pressure in the soil causes corresponding change in void ratio.

10. The change in thickness of the layer during consolidation in insignificant.
11. Constant values for certain soil properties remains constant.

Angle of internal friction

The measure of the resistance of soil to sliding along a plane is termed as angle of internal friction or friction angle ϕ. The angle of friction of soil mass depends upon—

1. Shape of the particles
2. Surface roughness
3. Type of interlocking
4. Lateral pressure
5. State of packing

Mohr's circle

Through every point within a mass subjected to a three dimensional stress system, three mutually perpendicular planes can be drawn, which experience only normal stress free from shear or tangent stress. These planes are called the principal planes and their normal stresses are called principal stresses.

Coulomb's law

According to Coulomb's law, the shearing strength of soil consists of the following two components—

1. Cohesion between particles
2. Friction between particles

He suggested the following straight line equation—

$$\tau = C + \sigma \tan \phi$$

where τ = Shear stress at the failure of soil
C = Cohesion intercept
σ = Normal stress on soil
ϕ = Friction angle

Rankine's theory

Rankine based his theory upon the following assumptions to calculate active earth pressure—

(a) The soil mass is homogeneous dry, cohesionless and semi-infinite.

(b) The ground surface is a plane which may be horizontal or inclined.

(c) The back of the wall is smooth and vertical.

(d) The wall yields about the base.

Limitations of Rankine's theory

(i) As the retaining walls are usually constructed of masonary or cement concrete, the back of the wall is never smooth and hence frictional forces develop.

(ii) On the assumption that wall back is smooth, the resultant pressure must act parallel to the surface.

(iii) The wall back may not always be vertical.

(iv) The retained soil may not always be cohesionless.

Table 1

S. No.	Soil Type	Value of f_s in KN/m²
1.	Silt and soft clay	5 to 20
2.	Loose sand	10 to 30
3.	Dense sand	25 to 50
4.	Dense gravel	40 to 80
5.	Very stiff clay	45 to 160

Table 2

S. No.	Type of Soil	Degree of Plasticity	Consistency Limit			Limits of P. I.
			Liquid Limit	Plastic Limit	Plasticit y Limit	
1.	Sand (Cohesionless)	Non-plastic	20	20	0	0
2.	Silt (Partly cohesive)	Low plastic	25	20	5	< 7
3.	Silty clay (Cohesive)	Medium plastic	40	25	15	> 7 < 17
4.	Clay (Cohesive)	High plastic	70	40	30	< 17

EXERCISE

1. The soil transported by flowing water is called—
(A) Aeolian soil (B) Marine soil
(C) Alluvial soil (D) Lacustrine soil

2. The soil transported by wind is called—
(A) Aeolian soil (B) Marine soil
(C) Alluvial soil (D) Lacustrine soil

3. Non-cohesive soil is—
(A) Sand (B) Silt
(C) Clay (D) Clay and silt

4. 'Loess' is silty clay formed by the action of—
(A) Water
(B) Glacier
(C) Wind
(D) Gravitational force

5. 'Talus' is the soil transported by—
(A) Wind
(B) Water
(C) Glacier
(D) Gravitational force

6. Soils are basically—
(A) Organic materials
(B) Inorganic materials
(C) Mineral materials
(D) Organic and inorganic materials

7. Which of the following soils has the finest grains ?
(A) Sands (B) Silts
(C) Fine sands (D) Clays

8. Silts have the following property—
(A) Plasticity
(B) Limited plasticity
(C) No plasticity
(D) Elasticity

9. The minimum size of grains of silt soil is about—
(A) 0·01 mm (B) 0·3 mm
(C) 0·06 mm (D) 0·002 mm

10. The maximum size of grains of silt soil is about—
(A) 0·06 mm (B) 0·1 mm
(C) 0·5 mm (D) 2 mm

11. The maximum size of clay particle is—
(A) 0·1 mm (B) 0·03 mm
(C) 0·002 mm (D) 0·0002 mm

12. The ratio between the total volume of voids and the total volume of solids is called—
(A) Porosity (B) Void fraction
(C) Void ratio (D) Solid ratio

13. As per I.S.S. the specific gravity of soil is determined at—
(A) 10°C (B) 17°C
(C) 27°C (D) 47°C

14. When the pores of a soil are full of water then the soil will be called—
(A) Moist soil
(B) Fully saturated soil
(C) Plastic soil
(D) Hydrated soil

15. The ratio of volume of air voids to the volume of voids is called—
(A) Void density (B) Air content
(C) Porosity (D) None of these

16. The dry density of which sample is expected to be highest ?
(A) Organic clay (B) Bentonite
(C) Stiff clay (D) Dense sand

17. The dry density of which sample may be expected to be least ?
(A) Stiff clay (B) Bentonite
(C) Dense sand (D) Loose sand

18. The weight of water per unit of volume of water is called—
(A) Moisture content
(B) Density
(C) Unit weight of water
(D) Degree of saturation

19. The specific gravity of sandy soils is—
(A) 1·5 (B) 2·0
(C) 2·2 (D) 2·6

20. The size of aperture of I.S. Sieve 300 micron is—
(A) 0·0300 mm (B) 0·300 mm
(C) 0·297 mm (D) 0·305 mm

21. Sieve analysis of a soil sample is done if the particle do not pass through square openings of—
(A) 0·075 mm (B) 0·150 mm
(C) 4·75 mm (D) 0·300 mm

22. Coarse sieve analysis is done for that material which retains on the I.S. sieve of opening size—
(A) 7·5 mm (B) 4·75 mm
(C) 2 mm (D) 1·50 micron

23. The effective size of a soil is—
(A) D_{10} (B) D_{20}
(C) D_{30} (D) D_{40}

24. A soil having uniformity coefficient more than 10, is called—
(A) Uniform soil (B) Poor soil
(C) Well graded soil (D) Coarse soil

25. A soil having uniformity coefficient more than 10 is called—
(A) Uniform (B) Fine
(C) Coarse (D) Well graded soil

26. The particle diameter corresponding to 20 percent finer on the grain size curve is called—
(A) D_{120} (B) D_{80}
(C) D_{20} (D) D_{200}

27. The minimum moisture content at which the soil just begins to crumple when rolled into 3 mm threads is known as—
(A) Elastic limit (B) Plastic limit
(C) Consistency limit (D) Liquid limit

28. The plastic limit exists in case of—
(A) Clays (B) Silty soils
(C) Sandy soils (D) Gravel soils

29. The percentage of the porosity of loose sand is in the range of—
(A) 5 to 15% (B) 15 to 30%
(C) 30 to 40% (D) 40 to 50%

30. The percentage of the porosity of compacted sand is about—
(A) 5 to 15% (B) 15 to 30%
(C) 30 to 40% (D) 40 to 50%

31. The shear strength of a soil in the plastic limit state is—
(A) Zero
(B) Reasonable
(C) Small
(D) Close to saturated soil strength

32. The shear strength of a soil in the liquid limit state is—
(A) Zero
(B) Very small
(C) Reasonable
(D) Close to saturated soil strength

33. The liquid and plastic limits exist in—
(A) Sandy soils (B) Silty soils
(C) Gravel soils (D) Clay soils

34. When the plastic limit of a soil is greater than the liquid limit, then the plasticity index is—
(A) Negative (B) Zero
(C) One (D) More than one

35. When the plasticity index of a soil is zero, the soil is—
(A) Clay (B) Silt
(C) Sand (D) Silty sand

36. The plasticity index of loose sand is about—
(A) 20–40 (B) 40–70
(C) 70–90 (D) None of these

37. The plastic index of highly plastic soil is about—
(A) 5–10 (B) 10–20
(C) 20–40 (D) More than 40

38. Permeability of soil is affected by—
(A) Grain size
(B) Void ratio of the soil
(C) Structural arrangement of the soil particles
(D) All the above

39. The dimensions of coefficient of permeability are—
(A) cm (B) g/cm^2
(C) cm/sec (D) g/cm^2

40. If the hydraulic gradient is unity, then the ratio of flow across unit area of soil is called—
(A) Coefficient of seepage
(B) Coefficient of permeability
(C) Coefficient of viscosity
(D) Coefficient of discharge

41. When a soil has coefficient of permeability as 1 mm/sec, then the soil should be—
(A) Clay (B) Silt
(C) Sand (D) Gravel

42. When a soil has coefficient of permeability as 0·00001 cm/sec, then the soil should be—
(A) Clay (B) Silt
(C) Sand (D) Gravel

43. The densification of a soil by machines is called—
(A) Consolidation (B) Compression
(C) Compaction (D) Soil stabilisation

44. The angle of internal friction of round grained loose sand is about—
(A) 5° to 25° (B) 25° to 30°
(C) 30° to 35° (D) 32° to 37°

45. The angle of internal friction of round grained dense sand is about—
(A) 5° to 25° (B) 25° to 30°
(C) 30° to 35° (D) 32° to 37°

46. The expansion of soil due to shear at a constant value of pressure is called—
(A) Apparent cohesion
(B) True cohesion
(C) Dialatancy
(D) Consistency

47. Surface tension is expressed in—
(A) Newtons
(B) Newtons per millimetre length
(C) Newtons per square millimetre
(D) Newtons per cubic millimetre

48. According to the Stoke's law, the terminal velocity depends on—
(A) Shape of the grains
(B) Size of the grains
(C) Weight of the grains
(D) All of the above

49. The seepage force in soils is proportional to—
(A) Exit gradient
(B) Head of water at upstream
(C) Head of water at downstream
(D) All of the above

50. A grillage foundation is essentially a—
(A) Shallow foundation
(B) Deep foundation
(C) Spread foundation
(D) Pile foundation

51. Foundations for buildings on sandy soils must have a depth of atleast—
(A) 0·1 m (B) 0·5 m
(C) 0·8 to 1m (D) 2 to 5 m

52. Maximum bearing capacity can be expected from—
(A) Laminated rocks
(B) Compact coarse sand
(C) Granite rocks
(D) Soft rocks

53. Which soil is expected to have least bearing capacity ?
(A) Laminated rocks
(B) Compact dry medium sand
(C) Loose fine sand
(D) None of these

54. A cohesive soil on drying—
(A) Expands (B) Shrinks
(C) Swells (D) Changes colour

55. Least bearing capacity may be expected from—
(A) Loose gravel
(B) Stiff clay
(C) Loose fine sand
(D) Compacted coarse sand

56. The shape of precast concrete pile is generally—
(A) Square
(B) Round
(C) Square or octagonal
(D) Square or circular

57. The bearing capacity of soil is influenced by the foundation with its—
(A) Shape
(B) Type

(C) Depth below ground
(D) Above all

58. Isolation of machine foundation is—
(A) Undesired (B) Desired
(C) Both (A) and (B) (D) Not predicable

59. Plate loading test is of—
(A) Short duration
(B) Long duration
(C) Medium duration
(D) Very long duration

60. The shape of test piece is—
(A) Hexagonal (B) Elliptical
(C) Rectangular (D) Square

61. Minimum size of the square plate has to be of—
(A) 10 cm side (B) 30 cm side
(C) 20 cm side (D) 50 cm side

62. The preliminary seating load to be applied as per I.S. is—
(A) 70 kgf/cm^2 (B) 70 kgf/mm^2
(C) 70 kgf/mm^2 (D) 70 kgf/cm^2

63. Test not recommended for design of foundation on clayey soils is—
(A) Plate test
(B) Standard penetration test
(C) Both (A) and (B)
(D) None of the above

64. Bearing capacity should be calculated from the criteria of—
(A) Shear only
(B) Settlement only
(C) Shear and settlement both
(D) All of the above

65. Prandtl's theoretical analysis assumes the soil as—
(A) Weightless (B) Isotropic
(C) Homogeneous (D) All the above

66. The angle of internal friction ϕ for cohesive soils is equal to—
(A) Zero (B) 30°
(C) 45° (D) 15°

67. The ultimate bearing capacity increases with increase in the—
(A) Depth of footing (B) Width
(C) Both (A) and (B) (D) None of these

68. Which type of foundation is preferable on soil of poor bearing capacity ?
(A) Raft (B) Stepped
(C) Grillage (D) None of these

69. Floating foundation is a type of—
(A) Well foundation (B) Raft foundation
(C) Pile foundation (D) None of these

70. The contact pressure, on uniformly loaded circular footing resting on clay, at the centre is—
(A) Minimum (B) Maximum
(C) Equal (D) Sinusoidal

71. Example of spread foundation is—
(A) Cantilever footing
(B) Inverted arch footing
(C) Pier foundation
(D) Both (A) and (B)

72. In the region of made-up-soil the type of foundation most suitable is—
(A) Isolated footing
(B) Pile foundation
(C) Combined footing
(D) Stepped foundation

73. Inverted arch footing is more suitable for—
(A) Residential buildings
(B) Bridges
(C) Single storey building
(D) Light warehouse

74. The height of capillary rise is maximum in—
(A) Fine gravel (B) Colloids
(C) Silt (D) Fine sand

75. The bearing capacity of a soil may be improved by—
(A) Reducing the depth of foundation
(B) Increasing the depth of foundation
(C) Moistening the soil
(D) Loosening the soil

76. The number of piles required to support a column is—
(A) 1 (B) 2
(C) 3 (D) 4

77. Pile foundations are generally used for—
(A) Runways
(B) Footpaths

(C) Bridges
(D) Sky scrapper buildings

78. Example of cased pile is—
(A) Swage pile
(B) Vibro pile
(C) Raymond pile
(D) Both (A) and (C)

79. Examples of load bearing piles may be—
(A) Sand piles (B) Composite piles
(C) Both (A) and (B) (D) None of these

80. The pressure piles are—
(A) Uncased (B) Cased
(C) Precast (D) All of these

81. When hard strata is not available at sufficient depth, the use is made of—
(A) Sheet pile (B) Friction pile
(C) Bearing pile (D) Battered pile

82. C. I. piles are not suitable for—
(A) Vibration resistance
(B) Heavy vertical loads
(C) Both (A) and (B)
(D) None of these

83. H-piles can resist—
(A) Earthquake forces
(B) Large horizontal forces
(C) Both (A) and (B)
(D) None of these

84. Steel H-piles are very much suitable for—
(A) Retaining walls
(B) Residential buildings
(C) School building
(D) All the above

85. Piers in pier foundation are built by—
(A) Excavation
(B) Hammer blows
(C) Vibro-equipments
(D) Hydro-forces

86. Minimum depth of embedment below scour level for piers should not be less than—
(A) 1·0 m (B) 2·0 m
(C) 0·5 m (D) 10·0 m

87. The consistency of a saturated cohesive soil is affected by—
(A) Water content
(B) Particle size distribution
(C) Both (A) and (B)
(D) None of these

88. A soil having particles of nearly the same size is known as—
(A) Well graded
(B) Uniformly graded
(C) Poorly graded
(D) Gap graded

89. Chances of settlement is more in—
(A) Cohesive soil
(B) Non-cohesive soil
(C) Both (A) and (B)
(D) None of these

90. Maximum permissible settlement for isolated foundation or sand as per I.S. code is—
(A) 20 mm (B) 10 mm
(C) 40 mm (D) 60 mm

ANSWERS

1. (C)	2. (A)	3. (A)	4. (C)	5. (D)
6. (D)	7. (D)	8. (C)	9. (D)	10. (A)
11. (C)	12. (C)	13. (C)	14. (B)	15. (B)
16. (D)	17. (B)	18. (C)	19. (D)	20. (A)
21. (A)	22. (C)	23. (A)	24. (C)	25. (D)
26. (C)	27. (B)	28. (B)	29. (D)	30. (C)
31. (C)	32. (B)	33. (D)	34. (B)	35. (C)
36. (D)	37. (C)	38. (D)	39. (C)	40. (B)
41. (C)	42. (A)	43. (C)	44. (B)	45. (D)
46. (C)	47. (B)	48. (D)	49. (A)	50. (C)
51. (C)	52. (C)	53. (C)	54. (B)	55. (C)
56. (C)	57. (D)	58. (B)	59. (A)	60. (D)
61. (B)	62. (D)	63. (B)	64. (C)	65. (D)
66. (A)	67. (C)	68. (A)	69. (B)	70. (A)
71. (D)	72. (B)	73. (B)	74. (B)	75. (B)
76. (C)	77. (D)	78. (D)	79. (C)	80. (A)
81. (B)	82. (A)	83. (B)	84. (A)	85. (A)
86. (B)	87. (A)	88. (B)	89. (A)	90. (C)

Chapter-7

THEORY OF STRUCTURES

Structural Members

Various types of structural members are used in frame structures. They are—

(i) **Beam**—The member which is subjected to bending is called beam. Beams are usually placed in horizontal position and are loaded with vertical loads.

(ii) **Tie**—The member which is subjected to only axial tension is called tie.

(iii) **Strut**—The member which is subjected to only axial compression is called a strut.

(iv) **Truss**—The group of ties and struts so designed and connected that these form a structure which acts as a large beam is called a truss.

(v) **Rigid frame**—The structure which moment resisting joints is called a rigid frame.

Loads

Various types of loads are applied to structural members. They are—

(i) **Dead load**—The load of constant magnitude that remains in one position is called dead load.

(ii) **Live load**—The load which does not remain in one position and may change in magnitude is called live load.

(iii) **Impact load**—The load which is caused by the vibrations of moving vehicles or movable loads is called impact load.

(iv) **Lateral load**—The load caused by the wind and earthquake is called lateral load.

(v) **Longitudinal load**—The load which is caused either by stopping a train or a railway bridge or a truck or a highway bridge is called a longitudinal load.

Support

(i) **Hinge support**—The support which is connected to the structure with a frictionless pin, is known as a hinge support. This type of support prevents both horizontal and vertical movements but is does not prevent even slight rotation about the hinge.

(ii) **Roller support**—The support which is provided with rollers is called roller support. This type of support prevents the movements only in a direction perpendicular to the supporting surface under the rollers.

(iii) **Fixed end support**—The support which provides resistence to rotation about the support and to the movement in vertical and horizontal directions is called a fixed end support.

Tabulated form of methods of analysis of various types of structures

S. No.	Type of Structure	Method of Analysis
1.	Determinate structures like beam, can-tilever simply supported, portal frames	Statical
2.	One-dimensional structures like beam- column, shaft, beamcolumn.	Statical if determine or else method of indeterminate structures.
3.	Two-dimensional structures like roof, floor, highways, domes, spherical and cylindrical shells.	Statical if determine or else method of indeterminate structure.
4.	Externally indeterminate structures like continuous beam, portal frame, fixed beam.	Strain energy method, 3 moment method, slope deflection method, moment distribution method.
5.	Internally indeterminate structure like trusses	Method of joints, method of sections.

Types of Structures

(i) **Statically determinate structure**—If the forces in the members of a structure and also its support reactions can be determined by the three conditions of equilibrium, *i.e.*,

$$\Sigma H = 0, \Sigma V = 0 \text{ and } \Sigma M = 0$$

(ii) **Statically indeterminate structure**—If the forces in the members of a structure and also its support reactions can not be determined by the three conditions of equilibrium, *i.e.*,

$$\Sigma H = 0, \Sigma V = 0 \text{ and } \Sigma M = 0$$

Castigliano's Theorems

Castigliano presented two theorems—

(i) **First theorem**—It states that "the partial derivative of the strain energy stored due to external load at a certain point of a statically determinate structure, is equal to deflection in the direction of the load at the point of its application."

(ii) **Second theorem**—It states that "In any and every case of statical indeterminate wherein, an indefinite number of different values of the redundant forces satisfy the conditions of statical equilibrium, their values are those that render total strain energy stored in the system to a minimum."

Strain Energy Method

When structural member is strained to external imposed loads, it stores some energy. This energy is called strain energy. The strain energy may be one of the following types—

(a) Strain energy due to direct stress

(b) Strain energy due to bending stress

(c) Strain energy due to shear stress

Analysis of Trusses

Various types of roof trusses are—

(a) **Warren and Pratt trusses**—These trusses are commonly used for flatter roofs for spans of roughly 12m to 38 m.

(b) **The pitched Pratt trusses**—These trusses are the most common types of medium size trusses which have maximum economical spans of about 30m.

(c) **Fink truss**—This type of truss is used for steep roofs.

(d) **Box string truss**—This type of truss is used for curved roofs for spans upto 35m.

(e) **Steel arch truss**—This type of truss is used for spans over 30m.

Influence Lines

An influence line is a curve, the ordinates to which at any point equals the value of some particular function due to unit load acting at that point.

Theorem 1—To obtain maximum value of a function due to a single concentrated load, the load should be placed at the point where the ordinate to the influence line for that function is maximum.

Theorem 2—The value of a function due to the action of a singly concentrated load equals the product of the magnitude of the load and the ordinate to the influence line for the function measured at the point of application of load.

Theorem 3—To obtain the maximum value of a function due to uniformly distributed load, the load should be placed over all those portions of the structure for which the ordinates to the influence line for that function have the sign of the character of the function desired.

Plastic Analysis

Following definitions are necessary to understand the principle of plastic analysis—

(i) **Elastic limit**—The largest stress that a material can withstand without being permanently deformed is called the elastic limit.

(ii) **Yield point**—The stress at which there is an increase in the strain without a corresponding increase in stress is called yield point.

(iii) **Elastic strain**—The strain that occurs before the yield point is called elastic strain.

(iv) **Plastic strain**—The strain that occurs after the yield point with no increase in stress is called plastic strain.

(v) **Strain hardening**—The range in the plastic zone where additional stress is necessary to produce additional strain is called strain hardening.

(vi) **Yield moment**—The moment that will just produce the yield stress in the outermost fibre of the section is called yield moment.

(vii) **Load factor**—The ratio of collapse load to the safe working load is known as load factor.

Principle of plastic analysis

The main principle of plastic analysis is based on the distribution of stresses at certain points in a structure reach the yield point. According to this theory, those parts of the structure that have been stressed to the yield point can not resist additional stress, but they will yield to the amount required to transfer the stresses to other parts of the structure.

EXERCISE

1. Maxwell's reciprocal theorem in structural analysis can be applied in—
   (A) All elastic structures
   (B) Plastic structures
   (C) Symmetrical structures only
   (D) Prismatic element structures only
2. The method of virtual work in the analysis of structures results in—
   (A) Compatible deformations
   (B) Equilibrium of forces
   (C) Stress strain relations
   (D) None of these
3. Unequal settlements in the supports of a statically indeterminate structure develop—
   (A) Member forces
   (B) Reactions from supports
   (C) No reactions
   (D) Strains in some members only
4. The force method in structural analysis always ensures—
   (A) Equilibrium
   (B) Kinematically admissible forces
   (C) Equilibrium of forces
   (D) None of the above
5. The analysis of statically indeterminate structures by the unit load method is based on—
   (A) Consistent deformation
   (B) Stiffness method
   (C) Consistent force
   (D) None of these
6. The moment area theorems in the structural analysis fall in the category of—
   (A) Force method
   (B) Displacement method
   (C) Stiffness method
   (D) Iterative method
7. The three moment equation in structural analysis is basically a—
   (A) Stiffness method
   (B) Displacement method
   (C) Energy method
   (D) Flexibility method
8. In moment distribution method the sum of distribution factors of all the members meeting at any joint is always—
   (A) Zero (B) < 1
   (C) > 1 (D) $= 1$
9. The carry over factor for prismatic member with far end fixed is—
   (A) $-\frac{1}{2}$ (B) $\frac{1}{2}$
   (C) $\frac{1}{4}$ (D) $-\frac{1}{4}$
10. The absolute stiffness of a prismatic member with one end fixed is—
   (A) $\frac{2EI}{L}$ (B) $\frac{4EI}{L}$
   (C) $\frac{3EI}{L}$ (D) None of these
11. The absolute stiffness of a prismatic member with one end hinged is—
   (A) $\frac{2EI}{l}$ (B) $\frac{4EI}{l}$
   (C) $\frac{3EI}{l}$ (D) None of these
12. The fixed support in real beam becomes in the conjugate beam is—
   (A) Fixed support (B) Hinged support
   (C) Roller support (D) Free support
13. The method of moment distribution in structural analysis is—
   (A) An iterative method
   (B) An exact method
   (C) An approximate method
   (D) None of these

14. The moment distribution method in structural analysis can be treated as—
    (A) Force method
    (B) Displacement method
    (C) Flexibility method
    (D) None of these

15. The torsional stiffness of a member can be defined as—
    (A) Torque for unit moment
    (B) Torque for unit twist
    (C) Moment for unit twist
    (D) Torsion for unit twist

16. The stiffness method in structural analysis is also known as—
    (A) Unit load method
    (B) Consistent deformation method
    (C) Force method
    (D) Displacement method

17. The strain energy stored in simply supported beam of span 'l' and flexural rigidity EI due to a ceratral concentrated load W is—
    (A) $\frac{W^2l^2}{84EI}$ (B) $\frac{W^2l^2}{48EI}$
    (C) $\frac{W^2l^3}{98EI}$ (D) $\frac{W^2l^3}{96EI}$

18. The ratio between the stress produced in a bar by a sudden applicaiton of load on compared to the stress produced by the gradual application of same load is—
    (A) 1·5 (B) 2·0
    (C) 2·5 (D) 3·0

19. A propped cantilever of span I carries a uniformly distributed load of 'W' per unit run over its entire span. The value of prop reaction to keep the beam horizontal is—
    (A) $\frac{WI}{3}$ (B) $\frac{3}{8}$ WI
    (C) $\frac{WI}{2}$ (D) $\frac{5}{8}$ WI

20. The horizontal thrust due to rise in temperature in a semicircular two-hinged arch of radius R is proportional to—
    (A) R (B) R^2
    (C) $\frac{1}{R}$ (D) $\frac{1}{R^2}$

21. A simply supported beam of span l and flexual rigidity EI, carries a unit point load at its centre. The strain energy in the beam due to bending is—
    (A) $\frac{l^3}{48EI}$ (B) $\frac{l^3}{192EI}$
    (C) $\frac{l^3}{96\,EI}$ (D) $\frac{l^3}{16\,EI}$

22. The maximum deflection of a fixed beam carrying a central load W is equal to—
    (A) $\frac{WL^3}{48EI}$ (B) $\frac{WL^3}{96EI}$
    (C) $\frac{WL^3}{192EI}$ (D) $\frac{5}{384}\frac{WL^3}{EI}$

23. In plastic analysis the shape factor for a circular section is—
    (A) 1·5 (B) 1·6
    (C) 1·7 (D) 1·75

24. The truss can be analysed by the methods of joints when the number of unknown at a joint is equal to—
    (A) 1 (B) 2
    (C) 3 (D) 4

25. The maximum central deflection for a simply supported beam loaded with uniformly distributed load W is—
    (A) $\frac{5}{584}\frac{WL^4}{EI}$ (B) $\frac{5}{384}\cdot\frac{WL^4}{EI}$
    (C) $\frac{WL^3}{48\,EI}$ (D) $\frac{48\,WL^3}{EI}$

ANSWERS

1. (A)	2. (B)	3. (B)	4. (C)	5. (A)
6. (A)	7. (D)	8. (C)	9. (B)	10. (B)
11. (C)	12. (D)	13. (A)	14. (B)	15. (B)
16. (D)	17. (D)	18. (B)	19. (B)	20. (D)
21. (C)	22. (C)	23. (C)	24. (B)	25. (B)

●●●

Chapter-8

RAILWAY ENGINEERING

Survey

Surveying in railway projects is very important. The important types of survey are—

(i) Traffic survey
(ii) Reconnaissance survey
(iii) Preliminary or initial location survey
(iv) Detailed or final location survey

Permanent way

The combination of rails fitted on sleepers and resting on ballast and sub-grade is called a permanent way. The clear distance between inner faces of the rails is known as the gauge. Now-a-days following gauges are used in our country.

Types of Gauge	Gauge Width (Metres)
(i) Broad Gauge (B.G.)	1·676
(ii) Meter Gauge (M.G.)	1·0
(iii) Narrow Gauge (N.G.)	0·702
(iv) Feeder Track Gauge (L.G.)	0·010

Components of permanent way

Following are the components of a railway track.

(a) **Rail sections**—The rails provide a hard, smooth and unchanging surface for the passage of heavy moving loads. Rails are made of high carbon steel to withstand wear and tear. Flat footed rails are mostly used in railway track. The standard rail sections are the following :

Standard Rail Sections

Gauge	Type	Weight (kg/m)
Broad gauge	55 R	55
	45 R	45
	35 R	35
Metre gauge	35 R	35
	30 R	30
	25 R	25
Narrow gauge	25 R	25

(b) **Sleepers**—The support which keeps the rails apart at required distance, supports the rail and distribute the load to the ballast are called as sleepers. Sleepers are of different materials such as wood, steel, cast iron, R.C.C. and prestressed concrete.

(c) **Rail fastenings**—Rail fastenings and fixtures are used to connect the rails and sleepers together in their proper positions. The fixtures and fastening used in track fittings are—

(i) Fish plates
(ii) Spikes
(iii) Bolts
(iv) Chairs
(v) Blocks
(vi) Keys
(vii) Plates

(d) **Ballast**—It is broken stone placed or packed below the sleepers to transmit load from sleeper to the formation and at the same time allowing drainage of the track.

(e) **Subgrade and embankment**—Subgrade is the normally occurring ground which when prepared to receive the blast and track is called formation.

Creep of rails

The longitudinal movement of rails or a permanent way caused due to speedy movement of rolling stock is known as creep of rails. The principal causes of creep are—

(i) At the time of starting of trains, the wheels push the rails backward and during stopping operations, the wheels push the rail forward.

(ii) Due to movements of the wheel loads on the rails, the rail deflects as continuous beam and thus crests are formed near the supports.

(iii) As a result of change in temperature unequal expansion and contraction of rails leads to creep.

(iv) Creep is more on curves than on straight lengths.

(v) Creep develops more in the direction of down grade due to gravity.

(vi) Creep develops more in the direction of loaded trains.

Points and crossings

Point and crossing is an arrangement provided in the railway track to provide diverging arrangement for the train from one track to another. The points and crossings are :

(i) Cross-overs
(ii) Scissors cross overs
(iii) Slips
(iv) Fixed points
(v) Three throws
(vi) Double turn-outs
(vii) Gathering lines or ladder tracks
(viii) Gantlet track
(ix) Double junction

Component of turnout :

(i) Stock rails
(ii) Tongue rails
(iii) Lead rails
(iv) Heel block
(v) Point rail
(vi) Splice rail
(vii) Wing rails
(viii) Check rails

Station and Yards

Station

At a station on railway line traffic is booked and authority to proceed is given to the trains. These are the following types :

(i) Flag station is not meant to control the trains movement. It deals with traffic only.

(ii) Block station deals with control of movement of the train without dealing the straffic.

STATION may be further classified into three categories :

(i) Wayside station
(ii) Junction
(iii) Terminals

Yard

It is a system of tracks laid within definite limits for various purposes such as storing of vehicles, making up trains, despatched the vehicles etc. Yards may be classified as—

(i) Passenger yard facilitates movement of passengers.

(ii) In good yards the goods are received loaded and unloaded.

(iii) Marshalling yard. It stores and distributes empty as will as loaded wagons, as and when received from other stations.

(iv) Locomotive yard. It is meant for maintenance, repairing and overhauling of locomotives.

Signalling

Safety to the traffic movement on rails may be ascertained by signalling. It facilitates smooth traffic flow and shunting operations. It may be classified as—

(i) Semaphore type signals
(ii) Warner signals
(iii) Disc or ground signals
(iv) Coloured light signals

Interlocking

The mechanical systems, levers and cables etc. which are essential in the operation of signal mechanism are known as interlocking. The methods for interlocking are—

(i) Tappets and locks systems
(ii) Key system
(iii) Route relay system

Key system is the simplest method of interlocking.

Maintenance of Track

(a) **Gauge maintenance**—Uniformity of the gauge should be properly maintained. Gauge may be disturbed due to following reasons.

(i) The loosening of track fittings

(ii) The irregularity of the tract gauge increases with the passage of time.

(b) **Joints maintenance**—A joint in the railway track is the weakest part. Proper attention should be paid for its maintenance. Riding, blowing and pumping are the defects which occur due to faulty maintenance of the rail joints and its surface.

(c) **Packing of ballast**—Due to heavy loads and vibrations caused by running trains, the ballast under the sleepers gets loose.

(d) **Track drainage**—The sources of moisture in a railway track are surface water, seepage water and hydroscopic water. The pressure of excessive water reduces the track stability erodes the banks of the embankments and in some cases may even result in accidents.

Station Machinery and Equipment

At junctions and terminal stations among others following machineries are generally provided—

(i) **Engine shed**—It is used for maintenance, repair and upkeeping of locomotives.

(ii) **Ash-pits**—These help in collection of locomotive's ash.

(iii) **Examination pits**—These are long and deep pits in locomotive yards to examine the engines from below.

(iv) **Water column**—These are provided for water supply to the locomotive on main lines.

(v) **Triangles**—These help in changing the direction of engine.

(vi) **Turn tables**—For change in direction of engine.

(vii) **Traversers**—Help for sideway shifting.

(viii) **Cranes**—Loading and unloading of vehicles is done by then.

EXERCISE

1. Generally the rail section used India are—
   (A) Double headed
   (B) Bull headed
   (C) Flat footed
   (D) All the above are correct

2. The choice of gauge depends upon—
   (A) Volume and nature of traffic
   (B) Speed of train
   (C) Physical features of the country
   (D) All the above

3. For connecting main cities and for routes of importance the type of gauge adopted is—
   (A) Meter gauge (B) Broad gauge
   (C) Narrow gauge (D) Mono rail

4. First train was run in India in April 1853 between—
   (A) Bombay-Thane
   (B) Delhi-Meerut
   (C) Madras-Pondicherry
   (D) Kolkata-Hawrah

5. Indian railway system is divided into—
   (A) Six zones (B) Nine zones
   (C) Three zones (D) Eight zones

6. As compared to roadways maintenance cost of railways is—
   (A) Same (B) More
   (C) Less (D) Depends on region

7. Majority of permanent way length is under—
   (A) Meter gauge (B) Narrow gauge
   (C) Broad gauge (D) Standard gauge

8. In India, the ratio between the weight of the rail and the locomotive axle load is—
   (A) 310 (B) 410
   (C) 510 (D) 610

9. Head width of 52 kg rail section is—
   (A) 61·9 mm (B) 66·7 mm
   (C) 67 mm (D) 72·33 mm

10. The cross sectional area of a 52 kg flat footed rail is—
    (A) 61·55 cm^2 (B) 66·15 cm^2
    (C) 72·35 cm^2 (D) None of these

11. Weight of rail used for meter gauge is—
    (A) 24·8 to 29·8 kg (B) 30 to 33 kg
    (C) 44·7 kg (D) 52 kg

12. The rails are made of—
    (A) Cast iron
    (B) Mild steel
    (C) High Carbon steel
    (D) High speed steel

13. The rail is designed by its—
    (A) Weight
    (B) Length
    (C) Weight/unit length
    (D) Cross-section

14. Bending of rails to connect curvature is done if the curve is—
(A) Less than 3 (B) More than 3
(C) Less than 4 (D) More than 4

15. Maximum wheel base distance provided on Indian B.G. tracks is—
(A) 4·096 m (B) 6·096 m
(C) 7·096 m (D) 9·096 m

16. The tread of wheels is provided an outward slope of—
(A) 1 in 10 (B) 1 in 15
(C) 1 in 20 (D) 1 in 25

17. Top of rails of a track are placed at an inward slope of—
(A) 1 in 10 (B) 1 in 15
(C) 1 in 20 (D) 1 in 30

18. On Indian Railways the standard length of rails for meter gauge track is—
(A) 10·06 m (B) 10·95 m
(C) 11·98 m (D) 12·8 m

19. Height of the rail for 52 kg section rail is—
(A) 129 mm (B) 143 mm
(C) 156 mm (D) 172 mm

20. Maximum degree of curvature for meter gauge is limited to—
(A) 10° (B) 16°
(C) 30° (D) 40°

21. The compulsory test conducted for rails is—
(A) Hammer test (B) Falling weight test
(C) Tensile test (D) Both (A) and (B)

22. On sharp curves the distance between the inner rail and check rail is—
(A) 33 mm (B) 44 mm
(C) 48 mm (D) 55 mm

23. Bull headed rails are generally provided at—
(A) Curved tracks
(B) Points and crossings
(C) Bifurcation of tracks
(D) Bridges

24. Creep is the—
(A) Longitudinal movement of rail
(B) Lateral movement of rail
(C) Vertical movement of rail
(D) None of these

25. Anticreep bearing plates are provided on—
(A) Bridge
(B) Joints
(C) Both sides of sleepers
(D) None of the above

26. Creep of rails is measured by—
(A) Creep indicator (B) Fishing string
(C) Anchors (D) None of these

27. Usually adjustment of rails is needed when creep exceeds—
(A) 10 mm (B) 50 mm
(C) 150 mm (D) 100 mm

28. The rails get damaged due to—
(A) Wear
(B) Bending
(C) Excessive hogging defect
(D) All the above

29. The causes of formation of kinds in a rail is—
(A) Loose packing at joints
(B) Defect in cross level at joints
(C) Defect in gauge and alignment
(D) All of the above

30. The misalignment of rails due to temperature changes is known as—
(A) Hogging (B) Creeping
(C) Buckling (D) Bulging

31. When the rail end rests on a joint steeper, the joint is termed as—
(A) Supported joint (B) Suspended joint
(C) Base joint (D) Welded joint

32. Staggered joints are generally provided—
(A) On curves
(B) On straight tracks
(C) To join two rail sections
(D) None of these

33. Number of fish bolts per fish plate is—
(A) 2 (B) 4
(C) 6 (D) 8

34. The most commonly used type of metal key is—
(A) Stuart's key (B) Spring coiled key
(C) Morgan key (D) All the above

35. The thickness of fish plate generally used in Indian Railway is—
(A) 10 mm (B) 16 mm
(C) 20 mm (D) 25 mm

36. Generally the shape of fish plate is—
(A) Rectangular (B) Circular
(C) Bone shaped (D) Eliptical

37. Rails are welded by—
(A) Gas welding (B) Arc welding
(C) TIG welding (D) Thermit welding

38. Perfect and strongest type of joints is a—
(A) Welded joint
(B) Staggered joint
(C) Supported joint
(D) Suspended joint

39. Generally the life of wooden sleepers is taken as—
(A) 2 to 3 years (B) 5 to 8 years
(C) 12 to 15 years (D) 35 to 50 years

40. The minimum length of wooden sleepers used on B.G. line is—
(A) 2 m (B) 2·2 m
(C) 2·4 m (D) 2·6 m

41. Standard size of wooden sleepers for M.G. track is—
(A) 274 × 25 × 13 cm
(B) 183 × 20 × 11 cm
(C) 152 × 15 × 10 cm
(D) 250 × 20 × 12 cm

42. Sleepers preferred on joints are—
(A) Wooden sleepers
(B) Concrete sleepers
(C) Steel sleepers
(D) CST-9 sleepers

43. Best wood for wooden sleepers is—
(A) Chir (B) Sal
(C) Teak (D) Shesham

44. Ideal sleepers are—
(A) R.C.C. sleepers
(B) Steel sleepers
(C) Cast iron sleepers
(D) Wooden sleepers

45. The sleepers providing the best rigidity of track is—
(A) Wooden (B) R.C.C.
(C) Cast iron (D) All of the above

46. Loose jaws of steel through sleepers are made of—
(A) Cast steel (B) Cast iron
(C) Spring steel (D) Mild steel

47. The formation width for a single line meter gauge track in embankment is—
(A) 3·70 m (B) 4·88 m
(C) 6·1 m (D) 5·49 m

48. The formation width for double line B.G. track is cutting excluding side drains takes as—
(A) 7·01 m (B) 7·93 m
(C) 10·06 m (D) 10·82 m

49. The side slope of embankment for a railway track is taken as—
(A) 1 : 1 (B) 1·5 : 1
(C) 2 : 1 (D) 2·5 : 1

50. Dog spikes are used for fixing rail to the—
(A) Wooden sleepers
(B) Concrete sleepers
(C) CST-9 sleepers
(D) Steel sleepers

51. Overall depth of a dog spike is—
(A) 120·6 mm (B) 135 mm
(C) 155·90 mm (D) None of these

52. The spike used for fixing chairs of bull headed rail to wooden sleepers is—
(A) Dog spike (B) Round spike
(C) Elastic spike (D) Rail spike

53. The sleepers resting directly on girder are fastened to the top flange of girder by—
(A) Dog bolts (B) Fish bolts
(C) Fang bolts (D) None of these

54. The normal size of the ballast used for points and crossings is—
(A) 10 mm (B) 25 mm
(C) 40 mm (D) 50 mm

55. The depth of ballast for a B.G. line is—
(A) 15-20 cm (B) 20-25 cm
(C) 25-30 cm (D) 35-40 cm

56. Sand is used as ballast in case of—
(A) Wooden sleepers
(B) Concrete sleepers
(C) Cast iron sleepers
(D) Steel sleepers

57. Two gradient generally adopted in India in plains is—
(A) 1 in 100 to 1 in 150
(B) 1 in 150 to 1 in 250
(C) 1 in 200 to 1 in 250
(D) 1 in 500

58. Minimum gradient in station yards is limited to—
(A) Zero (B) 1 in 500
(C) 1 in 750 (D) 1 in 1000

59. The shape of transition curve used by Indian railway is—
(A) Spiral (B) Sine curve
(C) Cubic parabola (D) Lemniscate

60. According to the Railway Board, the limiting value of cant is—
(A) $\frac{G}{5}$ (B) $\frac{G}{10}$
(C) $\frac{G}{15}$ (D) None of these
where G = Gauge

61. Normally maximum cant permissible on M.G. is—
(A) 60 mm (B) 75 mm
(C) 90 mm (D) 140 mm

62. The limiting value of cant gradient for all gauge is—
(A) 1 in 250 (B) 1 in 360
(C) 1 in 729 (D) 1 in 1000

63. A triangle used for turning the face of locomotives consists of—
(A) Three turnouts
(B) Three splits
(C) One turnout and two splits
(D) Two turnouts and one split

64. The plate laying is usually done in India by—
(A) Side method
(B) American method
(C) Telescopic method
(D) All the above

65. The point from where the laying of a new railway track starts is known as—
(A) Rail head (B) Base
(C) Material depot (D) Station

66. The horizontal distance from the material depot to the rail head is called—
(A) Lead (B) Lift
(C) Rail head (D) Site distance

67. The point up to which the new railway track laid, at any time is called—
(A) Rail head (B) Base
(C) Terminal (D) Station

68. The minimum height of embankment above the highest flood mark in the area should be—
(A) 30 cm (B) 60 cm
(C) Zero cm (D) 100 cm

69. The railway station at which a track line meets a main line is called—
(A) Way side station (B) Junction station
(C) Terminal station (D) Flag station

70. Disc signals are provided for—
(A) Passenger trains
(B) Goods trains
(C) Marshalling operations
(D) Shunting operations

71. Device for interlocking is—
(A) Detector (B) Treadle bar
(C) Point lock (D) All the above

72. The latest system of signalling introduced is—
(A) Automatic block system
(B) Section clear system
(C) Pilot guard system
(D) C.T.C. system

73. Accidents can be avoided by adopting—
(A) C.T.C. system (B) A.T.C. system
(C) Interlocking (D) Pilot guard system

74. The length of track under the control of a Ganger is—
(A) 1 to 2 km (B) 3 to 5 km
(C) 5 to 6 km (D) 10 to 15 km

75. The head of a gang in maintenance organisation of railways is called—
(A) A.P.W.I. (B) Ganger
(C) P.W.I. (D) Keyman

76. One degree of curve is equal to—
(A) 1700/R (B) 1720/R
(C) 1750/R (D) None of these
Where R = radius of curve

ANSWERS

1. (C)	2. (D)	3. (B)	4. (A)	5. (B)
6. (C)	7. (C)	8. (C)	9. (C)	10. (B)
11. (A)	12. (A)	13. (C)	14. (D)	15. (B)
16. (C)	17. (C)	18. (C)	19. (C)	20. (B)
21. (B)	22. (B)	23. (B)	24. (A)	25. (C)
26. (A)	27. (C)	28. (D)	29. (D)	30. (C)
31. (A)	32. (A)	33. (B)	34. (A)	35. (C)
36. (C)	37. (D)	38. (A)	39. (C)	40. (C)
41. (B)	42. (A)	43. (C)	44. (D)	45. (B)
46. (C)	47. (B)	48. (C)	49. (C)	50. (A)
51. (A)	52. (B)	53. (A)	54. (B)	55. (D)
56. (C)	57. (B)	58. (A)	59. (C)	60. (B)
61. (C)	62. (C)	63. (D)	64. (C)	65. (B)
66. (A)	67. (A)	68. (B)	69. (B)	70. (D)
71. (D)	72. (D)	73. (C)	74. (C)	75. (B)
76. (C)				

Chapter-9

HIGHWAYS ENGINEERING

According to location and purpose roads are classified as follows—

(1) National highway

(2) Provincial or State highway

(3) District road—(i) Major district road

(ii) Minor district road

(4) Village road

According to mode of construction—

(1) Earthern road or kuchcha road

(2) Water macadam road

(3) Bituminous road or black top road

(4) Cement concrete road

(5) Gravel roads are muram roads

National highways are the important road and connecting capitals or important cities of different states. These may also connect the neighbouring countries. The construction and maintenance of these roads being done by C.P.W.D. or M.E.S. They are numbered as NH-1, NH-2 etc.

State highways are the important roads of a state connecting important cities and towns within the state or important cities and district head quarters with a national highways.

Road geometrics

The visible dimensions are technically known as road geometrics. Road geometrics can be divided as follows—

(i) Right of way or permanent land width

(ii) Road margins

(iii) Formation width

(iv) Carriage way width

(v) Shoulders or footpath

(vi) Side slopes

(vii) Kerbs

(viii) Formation level

(ix) Camber or cross fall or cross slope

(x) Gradient

(xi) Super elevation

(xii) Curves

(i) **Right of way**—It is the area of land acquired and reserved for construction and development of the road along its alignment.

Permanent Land width as per Recommendations of I.R.C.

Class of Road	Right of way widths in metres			
	Open country		Built up Area	
	Minimum	Desirable	Minimum	Desirable
National and state highways	30·5	30·5 to 61·0	24·4	24·4 to 61·0
Major district roads	24·4	24·4 to 30·5	12·2	12·2 to 18·3
Other district roads	12·2	12·2 to 24·4	9·2	9·2 to 15·3
Village roads	12·2	12·2 to 18.3	9·2	9·2 to 15·3

(ii) **Road margins**—It is the portion of land width on either side of formation width of a road. It includes footpaths, cycle track, guard raid, embankment slopes and parking lane etc.

(iii) **Formation width**—It is the top width of road embankment or bottom width of road cutting excluding the side drains. It is also known as crest.

Formation width as per Recommendation of I.R.C.

Class of Roads	Minimum formation width in Metres	
	Plain Area	Hilly Area or Rock cutting
National and state highways	12·0	7·90
Major district roads	10·0	6·70
Other district roads	8·0	6·70
Village roads	7·5	4·25

(iv) **Carriage way**—It is the portion of the roadways used by vehicular traffic. It is in general the metalled portion of the road as per Indian Road Congress (I.R.C.) it is 3·8 metre for single lane traffic and 3·5 m. per lane in case of two or more than two lanes.

(v) **Shoulder for footpath**—It is the portion of the roadway between the outer edge of the pavement and top edge of the side slope. Its width generally varies between 1·2 to 1·8 m.

(vi) **Side slopes**—The slopes given to the sides of earthwork of road in embankment or cutting for its stability are called side slopes.

(vii) **Kerbs**—The boundaries between the pavement and footpath is known as kerbs.

(viii) **Formation level**—The reduced level of the finished surface of the earth work for a roadway or railway in embankment or cutting is known as formation level.

(ix) **Camber**—The central highest point of a curved road surface is known as crown of the road.

As per I.R.C. Recommendations

Types of Roads Surface	Camber
Earth roads, foot paths	1 in 20 to 24
Gravel roads	1 in 24 to 30
Water bound macadam roads	1 in 30 to 48
Bituminous roads inside towns	1 in 30 to 48
Bituminous roads outside town paving	1 in 48 to 60
Cement concrete roads	1 in 60 to 72

(x) **Gradient**—Gradient is the longitudinal slope of the road. The slope of the road which is adopted in road design and calculations of cuts and fills is called 'Ruling Gradient'. It is of three types—

(1) Limiting gradient

(2) Exceptional gradient

(3) Minimum gradient

(xi) **Super elevation**—It is the inward tilt of the road surface provided at the horizontal curves to compensate the effect of centrifugal force.

(xii) **Curves**—The geometrical arcs provided at the intersection of straight alignments either in horizontal or in vertical plane to have gradual change in direction is known as a curve.

Highway Geometric Design

Widening of pavements on curves—The width of road on curves, is increased due to following reasons—

(i) More sight distance on the sides of a pavement is required for safe driving.

(ii) On the curves drives find a difficulty to steer the vehicles along the centre line of the lane.

(iii) For overtaking another vehicle, more clearance for the safety and sense of security is desired.

(iv) The rear wheels of a vehicle on the curved portion of a pavement do not follow the front wheels, thus the sides of the vehicle do not remain parallel to the pavement edges, but these remains inclined.

Traffic Capacity of Highways

(i) **Traffic volume**—The number of vehicles moving in a specified direction on a roadway that pass a given point during a specified unit of time is called traffic volume.

(ii) **Traffic density**—The number of vehicles occupying a unit length of a roadway at a given instant, is called traffic density.

(iii) **Traffic capacity**—The maximum number of vehicles in a roadway that can pass a given point in unit time is called traffic capacity.

(iv) **Basic capacity**—The maximum number of passengers cars that can pass a given point on roadway during one hour under the most ideal road and traffic conditions, is called basic capacity.

(v) **Practical capacity**—The maximum number of vehicles that can pass a given point on a roadway during one hour without imposing any restriction to the driver's freedom for driving the vehicles, under prevailing roadway and traffic conditions is called practical capacity.

Types of Roads

(i) **Earthen roads**—The centre line and road edges are marked on the ground along the alignment with the help of pegs. Then site is cleared by removing all the vegetables and the surface is brought to the required camber and gradient by depositing or excavating the earth.

(ii) **Gravel road**—It is a cheap type of road. Gravels are obtained from river beds and by crushing the stones. The size of gravel varies from 6 mm to 35 mm. This gravel is mixed with sand and clay which acts as a binder.

(iii) **Water bound macadam road**—It is used as village road or a base for other important high cost roads. Road metals are bound together with the help of water and smaller pieces of metal are used to fill up the voids.

(iv) **Bituminous road**—The layer of bitumen surface dressing is provided to eliminate dust nuissance, to increase life of the road and to make it impervious and smooth.

(v) **Cement concrete road**—On the base course a concrete slab is laid to a thickness of 5 to 8 cm. This type is most expensive but very durable.

Construction of Road

(i) **Subgrade**—It is the soil foundation on which the road pavement rests and provides adequate and uniform support to the pavement.

(ii) **Sub-base course**—It is a layer of granular material provided in between the subgrade and the base course in a road surface.

(iii) **Base course**—It is a layer of boulders or over burnt bricks provided over the sub-base course or over sub-grade, if sub-base course is not provided.

(iv) **Base coat**—It is the layer of hard stones provided in between the base course and the wearing course.

(v) **Wearing course**—It is the top most layer of a road pavement which is directly exposed to the traffic.

Construction Materials in Highways

The construction of highway requires use of a large variety of materials, some of the main materials used are—

(i) Earth	(ii) Sand
(iii) Silt	(iv) Clay
(v) Gravel	(vi) Stone
(vii) Cement	(viii) Bitumen
(ix) Tar	(x) Water
(xi) Asphalt	(xii) Brick

Road Aggregate Tests—

Name of Text	Equipments used	Specification
Crushing test	Compression testing machine	Aggregate crushing value should be less than 45
Abrasion test	Los Angeles abrasion testing machine	L.A. abrasion values should be less than 50
Impact test	Aggregate impact testing machine	Aggregate impact value should not exceed 40%
Flakiness index test	Length gauge and sieves	Elongation index should not exceed 15%
Angularity number test	Metallic scoop and cylinder and sieves	Angularity number should be upto 11
Water absorption test	Dipping water bucket and balance	Absorption should be less than 0·6%
Bitumen adhesion test	Static immersion test	Stripping value of aggregate should not exceed 25%

Various methods of design based on theoretical, empirical and semi-empirical basis for design are—

(i) CBR method
(ii) Group index method
(iii) Triaxial test method
(iv) Stabilometer method
(v) Burmister method
(vi) Mcleod method

Traffic signals

Traffic signals are the control devices which alternately direct the traffic to stop and proceed at intersections. The signals are classified into following three types—

(1) **The traffic control signals**—These are provided with three coloured light glows facing

each direction of traffic flow *viz*, red, yellow amber and green. It has three subtypes :

(i) Fixed time signals

(ii) Traffic actuated signals

(iii) Manually operated signals

(2) **Pedestrian signals**—When the vehicular traffic remains stopped by red or stop signal on the traffic signals of the road intersection, these signals give the right of way to pedestrians to cross a road during the walk period.

(3) **Flashing beacons**—These signals are used to warn the traffic. Flashing yellow signals are used to direct the drivers of the vehicular traffic to proceed with caution.

Airport Geometrics—

(i) Runway length
(ii) Runway width
(iii) Sight distance
(iv) Longitudinal grades and grade changes
(v) Transverse grade
(vi) Runway intersections
(vii) Minimum clearance

Assemblage of various works mentioned below from a harbour—

(i) Dock
(ii) Quay
(iii) Wharve
(iv) Mole
(v) Catwalk

EXERCISE

1. The Indian Road Congress came into existence in—
(A) 1927 (B) 1934
(C) 1943 (D) 1947

2. Motor vehicle Act was enacted in—
(A) 1934 (B) 1939
(C) 1947 (D) 1951

3. The road connecting capitals of states is known as—
(A) National highway
(B) Provincial highway
(C) State highway
(D) Express highway

4. The road connecting one town with another town is designated as—
(A) National highway
(B) State highway
(C) District road
(D) Village road

5. An access from a road to private property is known as—
(A) Fly over (B) By pass road
(C) Loop road (D) Drive way

6. A road connecting one town with another is called a—
(A) Highway (B) Main road
(C) Trunk road (D) Country road

7. A road within a town is called—
(A) Trunk road (B) Country road
(C) Carriage way (D) Street

8. A portion of the traffic way that is used by the pedestrians only is called—
(A) Foot path (B) Side walk
(C) Foot way (D) Any of the above

9. The roads connecting village with towns or cities are classified as—
(A) Major district road
(B) State highway
(C) Minor district road
(D) Village road

10. The rate of rise or fall of a road along its alignment is known as—
(A) Gradient (B) Camber
(C) Side slope (D) Super elevation

11. Camber in the road is provided for—
(A) Effective drainage
(B) Counteracting the centrifugal force
(C) Having proper sight distance
(D) All of the above

12. The recommended safe co-efficient of friction is—
(A) 1·5 (B) 0·15
(C) 1/15 (D) 15

13. The highest point on a carriage way is known as—
(A) Crown (B) Camber
(C) Super elevation (D) Gradient

14. As per I.R.C. the camber on cement concrete road should be—
(A) 1 in 60 to 72 (B) 1 in 45 to 60
(C) 1 in 20 to 24 (D) 1 in 12 to 16

15. The road length per hundred square kilometres in India is—
(A) 10–15 km (B) 35–50 km
(C) 100–150 km (D) 200–250 km

16. As Assistant Engineer directly controls—
(A) Junior Engineer
(B) Work Assistants
(C) Skilled
(D) Executive Engineer

17. As per Indian Road Congress, the slope of earth in cutting should be—
(A) 1 : 1 (B) 1 : 2
(C) 1 : 4 (D) 2 : 3

18. As per I.R.C. the maximum width of a vehicle should be—
(A) 2 m (B) 2·44 m
(C) 3·8 m (D) 1·58 m

19. The maximum allowable super elevation is—
(A) 1 in 12 (B) 1 in 18
(C) 1 in 15 (D) 1 in 30

20. As per I.R.C. the carriage way width for two lane traffic should be—
(A) 7·5 m (B) 10 m
(C) 12·5 m (D) 15 m

21. The convexity provided on the surface of a highway is known as—
(A) Camber (B) Gradient
(C) Super elevation (D) Curve

22. Minimum recommended road width in case of National Highway is—
(A) 15 m (B) 25 m
(C) 40 m (D) 50 m

23. The value of ruling gradient in plains as per IRC is—
(A) 1 in 10 (B) 1 in 15
(C) 1 in 20 (D) 1 in 30

24. The steepest gradient in ordinary condition should not exceed—
(A) Ruling gradient
(B) Maximum gradient
(C) Exceptional gradient
(D) Floating gradient

25. The slope of the line joining the crown and edge of the road surface is known as—
(A) Cross fall (B) Cross-slope
(C) Camber (D) Any of the above

26. The road surfacing should be—
(A) Impervious (B) Durable
(C) Stable (D) All of the above

27. The thickness of the base in no case should be more than—
(A) 10 cm (B) 15 cm
(C) 20 cm (D) 30 cm

28. The reduction in load carrying capacity at a gradient of 1 in 20 is—
(A) 5% (B) 10%
(C) 15% (D) 20%

29. The gradient at hair-pin bends or other sharp corners with inside curves of 10 to 15 m should never exceed—
(A) 1 in 100 (B) 1 in 60
(C) 1 in 20 (D) 1 in 10

30. The thickness of the road surfacing depends upon the—
(A) Type of traffic
(B) Intensity of traffic
(C) Type of material
(D) All of the above

31. Weight of vehicles affects—
(A) Passing sight distance
(B) Extra widening
(C) Pavement thickness
(D) Width of lanes

32. The alignment of a road should cross the series of hills through—
(A) Ridge (B) Saddle
(C) Valley (D) None of the above

33. In a proposed alignment the type of soil which will not be preferred is—
(A) Soft rock (B) Ordinary soil
(C) Hard rock (D) All are correct

34. Which test is performed for quick determination of the quality of subgrade soil ?
(A) C.B.R. (B) Stripping
(C) Thread (D) None of these

35. In earthen roads the common stabilizers used are—
(A) Cement
(B) Bitumen
(C) Lime
(D) All the above are correct

36. The distance between pedestrian crossing and the stop line at a signalised intersection will be—
(A) 2 to 3 m (B) 8 to 10 m
(C) 1 to 2 m (D) 50 to 60 m

37. Mandatory signs are displayed on a disc having a diameter of—
(A) 30 cm (B) 60 cm
(C) 50 cm (D) 100 cm

38. A road sign indicates 'No parking' is—
(A) Warning sign (B) Prohibitory sign
(C) Mandatory sign (D) Informatory sign

39. A road sign indicates 'Keep Left' is—
(A) Warning sign (B) Prohibitory sign
(C) Mandatory sign (D) Informatory sign

40. A road sign indicating 'Parking for four vehicles' is—
(A) Warning sign (B) Prohibitory sign
(C) Mandatory sign (D) Informatory sign

41. Earthen roads or low cost roads are generally constructed as—
(A) National highways
(B) State highways
(C) District roads
(D) Village roads

42. The camber and gradient on earthen roads should not be more than—
(A) 1 in 10 (B) 1 in 20
(C) 1 in 50 (D) 1 in 80

43. The cross slopes on earthen roads should not be less than—
(A) 1 in 10 (B) 1 in 20
(C) 1 in 25 (D) 1 in 50

44. Generally the life of an earthen road varies in between—
(A) 12 to 30 years (B) 5 to 10 years
(C) 2 to 4 years (D) Less than 1 year

45. For proper drainage, the earthen road should have a minimum gradient of—
(A) 1 in 50 (B) 1 in 80
(C) 1 in 100 (D) 1 in 120

46. In a gravel road, the building material used is—
(A) Cement (B) Lime
(C) Clay (D) Surkhi

47. In concrete roads, the course which can be ommitted is—
(A) Subgrade (B) Base course
(C) Base coat (D) Wearing course

48. In tack coat, bitumen is used at the rate of—
(A) 1 kg/m^2 (B) 0·5 kg/m^2
(C) 5 kg/m^2 (D) 10 kg/m^2

49. Generally, the premix carpet laid in India is of thickness—
(A) 15 cm (B) 10 cm
(C) 5 to 7·5 cm (D) 2 cm

50. Generally the thickness of asphaltic concrete adopted is—
(A) 2 cm (B) 3 cm
(C) 4 cm (D) 5 cm

51. The life of a bitumen road having average traffic of 1200 tonnes per day, may be expected to be—
(A) 1 year (B) 2–5 years
(C) 5–7 years (D) 10–15 years

52. The thickness of cement concrete slab generally adopted varies in between—
(A) 1 cm to 5 cm (B) 5 cm to 10 cm
(C) 10 cm to 15 cm (D) 15 cm to 20 cm

53. If a cement concrete road is properly constructed and maintained, then its expected life will be—
(A) One year (B) 10 years
(C) 20 years (D) 30 to 40 years

54. The width of bay on a concrete road is generally taken as—
(A) > 4·5 m (B) < 4·5 m
(C) 3 to 4 m (D) 2 to 3 m

55. Compaction of concrete slab is done with the help of—
(A) Light roller (B) Heavy roller
(C) Vibrator (D) Mixer

56. Bituminous grouted macadam roads may be expected to have a useful life of about—
(A) 2 to 3 years (B) 3 to 5 years
(C) 7 to 10 years (D) 15 to 20 years

57. The maximum super elevation on hill roads should not exceed—
(A) 7% (B) 8%
(C) 9% (D) 10%

58. The snow bound sections, the super elevation on roads should not be more than—
(A) 7% (B) 8%
(C) 9% (D) 10%

59. The thickness of the parapet wall, on the valley side of the road way, is usually kept as—
(A) 20 cm (B) 40 cm
(C) 60 cm (D) 80 cm

60. The height of the parapet wall is provided above the beam level is—
(A) 25 cm (B) 50 cm
(C) 75 cm (D) 100 cm

61. On a hill road, the radius of curve should not be less than—
(A) 100 m (B) 60 m
(C) 30 m (D) 15 m

62. In hill road minimum sight distance required is—
(A) Stopping sight distance
(B) Passing sight distance
(C) Braking distance
(D) None of these

63. The drains constructed on up slope of hill sides, are known as—
(A) Cross drains
(B) Under drains
(C) Catch water drains
(D) Side drains

64. Mango, Jamun, Gulmohar and Malina trees are suitable for—
(A) Loamy soil (B) Clayey soil
(C) Sandy soil (D) Usar soil

65. Neem, Imli, Dhak etc. are most suitable in—
(A) Loamy soil (B) Usar soil
(C) Clayey soil (D) Sandy soil

66. Normally trees are planted longitudinally at a distance of—
(A) 10 to 15 m (B) 5 to 10 m
(C) 2 to 5 m (D) 10 to 12 m

67. In cement concrete roads, the repair work of cracks should be done just at the end of—
(A) Rainy season (B) Summer season
(C) Spring (D) Winter season

68. As per I.S. recommendations the average illumination on important roads carrying fast moving traffic is—
(A) 5 lux (B) 10 lux
(C) 20 lux (D) 30 lux

69. The width of roads which can be lighted by mounting street lights on one side of the road is—
(A) 10 m (B) 20 m
(C) 25 m (D) 30 m

70. The colour of lights used for visibility during fog is—
(A) White (B) Red
(C) Yellow (D) Green

71. The ideal form the curve for the summit curve is—
(A) Spiral (B) Parabola
(C) Circle (D) Lemniscale

72. In using the data from a plate bearing test for determining the modulus of subgrade reaction, the value of settlement to be used is—
(A) 1·25 mm (B) 2·50 mm
(C) 3·75 mm (D) 1·75 mm

73. As per I.R.C. the recommended land width for National Highway passing through plain area is normally—
(A) 25 metre (B) 35 metre
(C) 45 metre (D) 60 metre

74. The result of ring and ball softening point test on asphalts is given in terms of—
(A) Viscosity (B) Time
(C) Flow (D) Temperature

75. An Enoscope is used for measuring—
(A) Running speed (B) Time mean speed
(C) Spot speed (D) Overall speed

76. Gravity model is used in transportation planning process for—
(A) Modal split (B) Trip distribution
(C) Trip generation (D) Trip assignment

77. The plasticity index of the fraction passing 425 micron sieve in case of sub base course should be—
(A) Less than 6 (B) Greater than 6
(C) Greater than 9 (D) Between 15 and 20

78. Ratio of width of the car parking area required at kerb for 30° parking relative to 60° parking is approximately—
(A) 0·3 (B) 0·7
(C) 0·8 (D) 2·0

79. Airport elevation is the reduced level about M.S.L. of—
(A) Control tower
(B) Highest point of the landing area
(C) Lowest point of the landing area
(D) None of these

80. From the end of an instrument runway, the approach surface rises outwards—
(A) 1 in 50 (B) 1 in 30
(C) 1 in 40 (D) 1 in 80

81. The strength of winds is measured with the help of—
(A) Benfort scale (B) Wind indicator
(C) Barometer (D) None of these

82. For class A Air port the difference of reduced levels of higher and lower edges of the control surface is—
(A) 25 m (B) 50 m
(C) 75 m (D) 100 m

83. The length of run way is increased per 300 m rise above M.S.L.—
(A) 3% (B) 4%
(C) 7% (D) 6%

84. For the taxiways the maximum longitudinal grades is—
(A) 1·5% (B) 3%
(C) 5% (D) 7%

85. For the taxiway the permissible rate of change of grade is—
(A) 1% (B) 3%
(C) 5% (D) 7%

86. Two single runways on an airport can be arranged in the form of—
(A) T shape (B) L shape
(C) X shape (D) Any of these

87. According to ICAO recommendations all markings on runway are painted—
(A) White (B) Yellow
(C) Black (D) Red

88. According to ICAO recommendations all markings on taxiways are painted—
(A) White (B) Blue
(C) Black (D) Yellow

89. The width of threshold markings is—
(A) 4 m (B) 10 m
(C) 15 m (D) 20 m

90. Runway lights are usually spaced at not more than—
(A) 60 m (B) 150 m
(C) 200 m (D) 250 m

91. Taxiway lighting colour is—
(A) Blue (B) Green
(C) Red (D) White

92. The pilot normally takes decision about landing when he is about—
(A) 100 m (B) 60 m
(C) 40 m (D) 20 m

93. The landing and take off area of heliports are marked with—
(A) Letter H (B) Letter P
(C) Number 3 (D) None of these

94. Holding aprons are also known as—
(A) Warm-up pads
(B) Run-up pads
(C) Both (A) and (B)
(D) None of these

95. Gates provided for dock entrances are known as—
    (A) Locks (B) Caissons
    (C) Dock wall (D) Open birth

96. Lowest tide of the mouth is known as—
    (A) Neap tide (B) Tidal bore
    (C) Spring tide (D) Diurnal tide

97. The returning current towards the sea is known as—
    (A) Flood current (B) Slack water
    (C) Ebb current (D) Mean current

98. Long duration water oscillations in the harbour is termed as—
    (A) Mean range
    (B) Sieches
    (C) Turn of the current
    (D) All the above

99. The tidal range is maximum in India at—
    (A) Mumbai (B) Chennai
    (C) Paradeep (D) Bhavnagar

100. The problem of water level difference in reaching the dock is overcommed by—
    (A) Lock (B) Weir
    (C) Orifice (D) Mouth piece

101. The lock gates are always in—
    (A) Single number (B) Pair
    (C) Quadruple (D) Dozen

ANSWERS

1. (B)	2. (B)	3. (A)	4. (B)	5. (D)
6. (A)	7. (D)	8. (D)	9. (D)	10. (A)
11. (A)	12. (B)	13. (A)	14. (A)	15. (B)
16. (A)	17. (A)	18. (B)	19. (B)	20. (A)
21. (A)	22. (D)	23. (D)	24. (B)	25. (D)
26. (D)	27. (D)	28. (B)	29. (C)	30. (D)
31. (C)	32. (B)	33. (C)	34. (A)	35. (D)
36. (B)	37. (B)	38. (C)	39. (C)	40. (D)
41. (D)	42. (B)	43. (D)	44. (C)	45. (D)
46. (C)	47. (C)	48. (B)	49. (D)	50. (C)
51. (D)	52. (D)	53. (D)	54. (B)	55. (C)
56. (C)	57. (D)	58. (A)	59. (C)	60. (C)
61. (D)	62. (A)	63. (C)	64. (D)	65. (D)
66. (A)	67. (B)	68. (D)	69. (A)	70. (C)
71. (B)	72. (A)	73. (C)	74. (D)	75. (C)
76. (B)	77. (A)	78. (A)	79. (B)	80. (A)
81. (A)	82. (D)	83. (C)	84. (B)	85. (A)
86. (D)	87. (A)	88. (D)	89. (A)	90. (A)
91. (A)	92. (B)	93. (A)	94. (C)	95. (B)
96. (A)	97. (C)	98. (B)	99. (D)	100. (A)
101. (B)				

Chapter-10

STEEL STRUCTURES

Steel Structure

Steel structure possess high strength. Due to this reason it gives the advantages of longer life, easy transportation, easily erected, assembled etc.

Loads carried by these structures are

Dead Load

(Self weight of the structure along with weight of all permanent fixtures attached to it.)

Live Load

(The load on a structure other than the dead load. For example—

(i) Superimposed loads

(ii) Dynamic loads

(iii) Wind loads

(iv) Snow loads

(v) Seismic loads

(vi) Temperature stresses

Steel

It is an alloy of iron, carbon and other elements in varying percentages. It is of two types—

(i) Mild steel

(ii) High tensile steel

Mild steel

Mild steel is used to manufacture rolled structural steel sections, rivets and bolts and is amenable to cutting, drilling, welding and machining etc.

High tensile steel

High tensile steel is used in the manufacture of cutting tools and where high strength is required.

Riveted connections

A small piece of round steel forged in place to connect two or more than two steel members together is known as rivet. It consists of one head and body. The body is named as shank.

Riveted joints

The riveted joints are classified as—

(i) **Lap joints**—In this the members to be connected together are made to overlap each other and then connected by riveting.

(ii) **Butt joints**—In this the members butt against each other and an over plate is used which is fastened to the main members by riveting.

Design of riveted joints

(i) **Minimum pitch**—The distance between centres of rivet holes should not be less than 2·5 times the diameter of the holes.

(ii) **Maximum pitch**—The distance between centres of any two adjacent rivets connecting together elements of compression and tension members should not exceed 32 t or 300 mm whichever is less where t = thickness of the thinner outside plate.

Welded Connections

The connection between two metal pieces by establishing a metallurgical bond between them by heat or by pressure or both is known as weld.

Main processes of welding

(1) **Fusion welding**—In fusion welding two metals are welded in a state of fusion without applying external pressure on the metals.

(2) Pressure welding

(3) Thermist welding

(4) Union Melt welding

Tension members

The members which are subjected to direct tension are called tension members, also known as tie members. Generally members used as tension members are—rod, round or squares, flats, eye bars, single angle sections, T-section and channel sections.

Types of Section	Thickness or Diameter	Tensile Strength in kg/cm²
Rolled 1 beam and channels	All and including upto	1500
Plates, bars, universal beams columns	20 mm > 20 mm	1500
Plates, section and bars	Upto and including 45 mm > 45 mm	2125 1810

Beams

Structural members usually straight and subjected to bending moments, shear forces due to transverse loads but carrying no axial thrust or tension are called beams.

Simple beam end connections may be of the following types—

(i) **Framed connections**—These may be double plate web and double angle web connections-welded or riveted. Beam preparation becomes economical at the cost of weld and plate materials.

(ii) **Seated connections**—These may be riveted or welded seat connections. For lateral stability of beam, top clear angle is used in unstiffened seat connection.

Types of beam—

(i)	Floor beam	(ii)	Joist
(iii)	Girder	(iv)	Rafter
(v)	Lintel	(vi)	Purlin
(vii)	Stringer	(viii)	Spandrel beam

Plate girder

When the span is large and load is heavy, rolled steel beams become insufficient to furnish the requirements and built up beams became uneconomical, then plate girders are used. Plate girders are deep structural members subjected to transverse loads. They are nothing but built up beams. They consist of plates and angles riveted together. Plates and angles form are I-section. They are used in building constructions and also in bridges. The plate girders are economically used for spans from 15 m to 30 m. The plate girders have been used for spans upto 40 m.

Plastic design of beams

The structural strength of a member depends upon the criterion of design adopted. In elastic design, the attainment of yield stress at any section of a member is assumed to limit its load carrying capacity whereas in plastic design failure of members due to fracture, fatigue and instability are considered to be premature failures and steps are taken to eliminate these so that a member may attain its maximum possible strength.

The following are two methods for determination of plastic moments—

(i) **Statical method**—It gives the lower bound of ultimate load. These loads calculated should satisfy $M \leq M_p$ condition. This is suitable for simple structures.

(ii) **Mechanism method**—This is convenient to solve structures with many degrees of redundancy. In these methods, possible plastic hinges are located.

Roof trusses

The truss is a frame work in which the members are connected at their ends. A series of trusses fixed on columns on the wall cover or floor area. The trusses support the purlins and purlins support at roof covering either directly or through common rafters.

Types of common roof truss—

(a) King post truss
(b) Queen post truss
(c) Howe-flat truss
(d) Howe-triangular truss
(e) Triangular truss
(f) North-light truss

Steel bridges

Bridges are made for various purposes such as for a highway, a railway track, for carrying pipe lines. While R.C.C. bridges are common in use for all the above purposes, steel bridges are more commonly used in Indian Railways for very small spans as well as very large spans.

Components of a bridge

In general following are components of a bridge structure—

(a) Foundations for the abutments and piers
(b) Abutment and piers

(c) River training works

(d) Approaches to the bridge

(e) Decking

(f) Bearings for the girders or trusses

(g) Hardrails

(h) Gurad stones etc.

Classification of bridges

Depending upon function to be performed, material of construction, position of floor, inter span relationship etc. bridges can be classified as follows—

(1) **Functional classifications**—Aqueduct, viaduct, highway, or foot, floating bridge.

(2) **Material of construction**—Timber, steel, R.C.C. bridges.

(3) **Position of floor**—Deck bridge, through, semithrough, or suspension bridge.

(4) **Inter span relationship**—Simple cantilever, continuous, arch type, suspended, rigid type etc.

(5) **Life of bridge**—Temporary, semi permanent and permanent bridge.

Types of bridges

The steel bridges may be classified on the basis of their use and mode of construction.

(1) **Deck bridge**—In this case the floor of bridge is located on the top flange of a plate girder.

(2) **Through bridge**—In this case, the floor is supported at lower chord level.

(3) **Pony truss bridge**—Here the floor is located at or near the bottom.

(4) **Double deck bridge**—In this case two decks such as road-cum-rail bridge are provided.

(5) **Plate girder bridge**—It is a built-up plate girder suitable for large span bridge.

(6) **Truss bridge**—It is economical for much longer span.

(7) **Balanced-cantilever bridge**—It is used for spans upto about 60 m.

Suitable Spans for Various Types of Steel Bridge

Types of Steel Bridge	Maximum Limit of Span in Metre
1. R.S. Joist	3 to 7·5
2. Steel through plate bridge	Upto 5
3. Girder bridge	7·5 to 18
4. Lattice girder bridge	18 to 30
5. Trussed bridge	30 to 80
6. Arch bridge	30 to 100
7. Bow string bridge	50 to 150
8. Suspension bridge	Above 150

Girder bridges

There are two types of girder bridges—

(a) Solid web girder

(b) Open web girder

EXERCISE

1. The density of steel used in the structural members should be—
   (A) 1 gm/mm^3 (B) 6·4 gm/cm^3
   (C) 7·9 gm/cm^3 (D) 13·6 gm/cm^3

2. A tie is a—
   (A) Flexible member
   (B) Compression member
   (C) Torsion member
   (D) Tension member

3. A strut is a—
   (A) Flexible member
   (B) Compression member
   (C) Torsion member
   (D) Tension member

4. The structures made of rigid curved surfaces are known as—
   (A) Surface structure
   (B) Frame structure
   (C) Shell structure
   (D) Space structure

5. Weight of steel per cubic metre is taken to be—
   (A) 7900 kg (B) 3000 kg
   (C) 1200 kg (D) 800 kg

6. Which operation can not be done easily on mild steel ?
   (A) Drilling (B) Punching
   (C) Cutting (D) Hardening

7. The minimum percentage elongation for mild steel is—
(A) 1 % (B) 2 %
(C) 12·3 % (D) 23 %

8. The percentage of carbon in structural steel is—
(A) 0·20 to 0·27 % (B) 0·70 to 0·80 %
(C) 1·20 to 1·27 % (D) 1·70 to 1·80 %

9. The percentage of sulphur in structural steel is restricted to—
(A) 0·055 (B) 0·15
(C) 0·55 (D) 0·75

10. Junction between the flange and web of a beam is known as—
(A) Lap joint (B) Butt joint
(C) Fillet (D) Shear joint

11. A channel section consist of
(A) Two webs
(B) Two flanges
(C) Two webs and two flanges
(D) One web and two flanges

12. A piece of round steel forged in place to connect two or more steel members is known as a—
(A) Bolt (B) Rivet
(C) Screw (D) Nail

13. The size of the rivet is expressed by the—
(A) Length of shank
(B) Diameter of shank
(C) Type of head
(D) None of these

14. The most commonly used rivet head is—
(A) Snap head (B) Pam head
(C) Flat head (D) All are correct

15. For structural steel works the diameter of hot driven rivets is—
(A) 16 mm
(B) 18 mm
(C) 20 mm to 22 mm
(D) All the above

16. The diameter of cold driven rivets ranges from—
(A) 6 mm to 12 mm
(B) 12 mm to 22 mm
(C) 22 mm to 32 mm
(D) 32 mm to 42 mm

17. Snap head of a rivet can also be termed as—
(A) Round head (B) Button head
(C) Both are wrong (D) Both are correct

18. For rivet diameter upto 25 mm the diameter of rivet hole is larger than the diameter of rivet by—
(A) 1 mm (B) 5 mm
(C) 2 mm (D) 2·5 mm

19. The maximum shear stress of steel member in flexure shall not exceed—
(A) 0·4 fy (B) 0·66 fy
(C) 0·5 fy (D) 0·45 fy

20. The maximum deflection for a simply supported beam should not exceed—
(A) $\frac{1}{325}$ of span (B) $\frac{1}{525}$ of span
(C) $\frac{1}{225}$ of span (D) None of these

21. As per ISI rolled steel beam sections are classified into—
(A) Two section (B) Three section
(C) Four section (D) Five section

22. The greatest grip of long rivets shall not exceed—
(A) 2 d (B) 4 d
(C) 8 d (D) 6 d

23. Rolled steel angle sections are classified—
(A) Equal angles (B) Unequal angles
(C) Bulb angles (D) All the above

24. If P is wind pressure is kg/cm^2, v = velocity in km/hr and K = constant of proportionality then—
(A) $P = K/v^2$ (B) $v = K/P^2$
(C) $P = Kv^2$ (D) None of these

25. An imaginary line along which rivets are placed is known as—
(A) Rivet line (B) Gauge line
(C) Back line (D) All of these

26. The greater clear dimension of web thickness 't' should not exceed—
(A) 180 t (B) 200 t
(C) 240 t (D) 270 t

27. The imaginary line along which rivets are to be placed is known as—
(A) Rivet line (B) Gauge line
(C) Scrieve line (D) Back line

28. Generally the diameter of rivets used in structural members is not less than—
(A) 6 mm (B) 12 mm
(C) 16 mm (D) 20 mm

29. Generally in bridges, rivets used are of the diameter ranging in between—
(A) 12 to 16 mm (B) 16 to 22 mm
(C) 22 to 25 mm (D) 25 to 40 mm

30. The metal added at the joint while welding is known as—
(A) Weld metal (B) Filler
(C) Fillet metal (D) All of these

31. The advantages of welding lies in—
(A) Quick and economic process
(B) Better appearance of joints
(C) Rigid joints
(D) All the above

32. As compared to riveting, welding is preferred these days because—
(A) It is a silent process
(B) It is more economical
(C) It has more efficient
(D) All the above

33. If a structure is under fatigue stresses then the welded joints as compared to riveted joints will fail—
(A) Earlier
(B) Later
(C) At the same time
(D) Not at all

34. As compared to welding, riveting is preferred due to—
(A) Ease in construction and maintenance
(B) Ease in the inspection of riveted joints
(C) Safety of the operator
(D) All are correct

35. A fillet weld is known as standard fillet weld if the cross-section of the fillet weld is a triangle of the angle of—
(A) 45° (B) 60°
(C) 30° (D) 75°

36. Fillet weld is not used for joining the parts, when the angle between fusion faces is—
(A) Greater than 120°
(B) Less than 60°
(C) Both are correct
(D) Both are wrong

37. For the structural members exposed to atmosphere, the most unsuitable type of weld is—
(A) Slot weld
(B) Butt weld
(C) Intermittent fillet weld
(D) Plug weld

38. The size of a butt weld is specified by—
(A) Throat thickness
(B) The effective throat thickness
(C) Thickness of thickest part joined
(D) None of the above

39. The effective throat thickness in mm of a 6 mm size fillet weld with an angle of 75° between the fusion faces is—
(A) 3·6 (B) 4·0
(C) 4·2 (D) 4·5

40. The size of deep penetration fillet weld is specified as minimum leg length plus—
(A) 1·2 mm (B) 2·4 mm
(C) 3·0 mm (D) 4·8 mm

41. The effective length of fillet should not be less than—
(A) The size of weld
(B) Two times the size of weld
(C) Three times the size of weld
(D) Four times the size of weld

42. The effective length of the intermillet weld, in respect of the thickness of the thinner part joined should not be less than—
(A) Two times (B) Four times
(C) Ten times (D) Sixteen times

43. The structural member subjected to compressive force in a direction parallel to its longitudinal axis is called—
(A) Post (B) Stanchion
(C) Column (D) Any one of these

44. When compression members are overloaded then their failure takes place because of—
(A) Direct compression
(B) Excessive bending
(C) Bending combined with twisting
(D) Any one of the above

45. The axial load on column which results in a slight deflected shape of column is called—
(A) Critical load (B) Crippling load
(C) Buckling load (D) Any of these

46. The failure of a column depends upon—
(A) Weight of column
(B) Length of column
(C) Slenderness ratio
(D) Cross sectional area of column

47. Slenderness ratio of the column will be—
(A) $l \times r$ (B) l/r
(C) $(l + r)^2$ (D) $\sqrt{l/r}$

48. In a structure generally wires and cables are used as—
(A) Compression member
(B) Tension member
(C) Compression or tension
(D) To resist shear stresses only

49. In a structure generally the members used as tension members are—
(A) Wires and cables
(B) Rods and bars
(C) Angle and tee section
(D) All are correct

50. The maximum value of slenderness ratio of compression members carrying loads resulting from dead loads and superimposed load is—
(A) 100 (B) 180
(C) 150 (D) 200

51. The most economical section for a column is—
(A) Square section
(B) Circular section
(C) Tubular section
(D) Hexagonal section

52. Effective length of a column is the length between the points of—
(A) Maximum moments
(B) Zero shear
(C) Zero moment
(D) None of these

53. In a structure generally wires and cables are used as—
(A) Compression member
(B) Tension member
(C) Compression or tension
(D) To resist shear stresses only

54. The main beam is a beam which supports—
(A) Floor construction
(B) Joists
(C) Secondary beam
(D) Both (B) and (C)

55. Any major beam in a structure is known as—
(A) Subsidiary beam
(B) Joist
(C) Girder
(D) Secondary beam

56. The lighter section of structural members subjected to transverse loading are called as—
(A) Struts (B) Tie
(C) Joists (D) Stanchions

57. In roof trusses the horizontal beam spanning between the two adjacent trusses are known as—
(A) Principal rafter (B) Common rafter
(C) Purlin (D) All of these

58. The beams at the outside wall of a building supporting its share of the floor and also the wall up to the floor above it are known as—
(A) Stringers (B) Spandrel beams
(C) Trimmers (D) Headers

59. In bridge floors the longitudinal beams supported by floor beams are also called as—
(A) Headers (B) Trimmers
(C) Stringers (D) Rafters

60. The beams resting on purlins are known as—
(A) Spandrel beams (B) Common rafters
(C) Trimmers (D) Stringers

61. Out of all the available rolled steel sections the most commonly used section as beam is—
(A) T-section (B) L-section
(C) I-section (D) Channel-section

62. As per I.S. 800–1971 the allowable bending stresses (compressive) in rolled I–beams and channels should be—
(A) 700 kg/cm^2 (B) 1650 kg/cm^2
(C) 1200 kg/cm^2 (D) 2000 kg/cm^2

63. The deep structural members subjected to transverse loads are called—
(A) Beam (B) Columns
(C) Plate girders (D) Trusses

64. The maximum span upto which plate girders can be used economically is—
(A) 15 m (B) 10 m
(C) 30 m (D) 40 m

65. The vertical plate of a plate girder is termed as—
(A) Web plate (B) Flange plate
(C) Cover plate (D) None of these

66. The girders having two or more than two webs are called—
(A) Plate girder (B) Box girder
(C) Gantry girder (D) All of these

67. The horizontal plates connected with the flange angle are known as—
(A) Web-plates
(B) Cover plates
(C) Flange plate
(D) Both (B) and (C) are correct

68. As compared to span, depth of plate girder is generally kept as—
(A) 1/4 to 1/5 (B) 1/8 to 1/10
(C) 1/10 to 1/12 (D) 1/20 to 1/24

69. Generally in a simply supported truss, the principal rafter will carry—
(A) Compressive ratio
(B) Tensile forces
(C) Sometimes compressive and sometimes tensile forces
(D) All the above are correct

70. Which of the following is necessary a natural defect in timber ?
(A) Knots
(B) Warping
(C) Repture of tissues
(D) All the above

71. Which of the following has influence on the strength of timber ?
(A) Moisture content
(B) Presence of defects
(C) Types of grains
(D) All the above

72. One cubic metre of teak wood weighs approximately—
(A) 200 kg (B) 625 kg
(C) 1250 kg (D) 1800 kg

73. Timber has more strength—
(A) Along the grains
(B) Across the grains
(C) Perpendicular to grains
(D) In all directions

74. The average unit weight of sal is 865 kg/m^3, provided the moisture content in the wood should be—
(A) 20 % (B) 12 %
(C) 6 % (D) Zero

75. In solid wood columns, the slenderness ratio should not exceed—
(A) 11 (B) 25
(C) 35 (D) 50

76. In a cantilever beam, the maximum deflections of the beam should not be greater than—
(A) $\frac{1}{180}$ (B) $\frac{1}{120}$
(C) $\frac{1}{360}$ (D) $\frac{1}{240}$

77. Which one of the following methods of design is not suitable for structures subjected to impact and fatigue ?
(A) Simple design
(B) Semi-rigid design
(C) Rigid design
(D) Plastic-design

78. Ties are load carrying members of a frame which are subjected to—
(A) Transverse loads
(B) Axial tension loads
(C) Axial compression loads
(D) Torsional loads

79. Generally the maximum deflection/span ratio of a steel member should not exceed—
(A) $\frac{1}{750}$ (B) $\frac{1}{500}$
(C) $\frac{1}{325}$ (D) $\frac{1}{250}$

80. The web buckling in beams may be avoided by—
(A) Stiffening the web
(B) Limiting the stress
(C) Increasing the span
(D) Any one of the above

81. The deflection of a beam may be reduced by—
(A) Decreasing the depth of beam
(B) Increasing the span
(C) Providing greater end restraint
(D) Any of the above

82. Size of rivet is specified by—
(A) Shape (B) Diameter of head
(C) Shank diameter (D) Weight

83. The shock absorbing capacity of a bolt is increased by—
(A) Proper tightening
(B) Use of spring washer
(C) Making shank diameter equal to the core diameters
(D) Preventing stress concentration anywhere in the bolt

84. As per IS 800–1984 the minimum pitch of rivets in a row is recommended as diameter of the rivet times—
(A) 2·0 (B) 2·5
(C) 3·0 (D) 4·0

85. The centre to centre distance between any two adjacent supports of a bridge is called—
(A) Total span (B) Gauge
(C) Span (D) Clear span

86. The bridge having its floor supported at the top structure is called—
(A) Through bridge
(B) Semi-through bridge
(C) Deck bridge
(D) Over bridge

87. Open bridge floor is suitable for—
(A) Railway bridge
(B) Roadway bridge
(C) Pedestrian bridge
(D) Navigable bridge

88. The T-beam bridge is generally constructed as—
(A) Deck bridge
(B) Through bridge
(C) Semi-through bridge
(D) None of these

89. The end-supports of a bridge are called—
(A) Piers (B) Abutments
(C) Wing walls (D) Abutment piers

90. The pair of walls constructed on both sides of abutments to retain the earth banks are called—
(A) Wing walls (B) Breast walls
(C) Retaining walls (D) Curtain walls

91. According to IRC recommendations the bridge approaches should be straight for a minimum length of—
(A) 25 m (B) 30 m
(C) 50 m (D) 15 m

92. The intermediate supports in a bridge are called—
(A) Bearing wall (B) Abutment
(C) Wing wall (D) Piers

93. Fixed plate bearings are suitable for spans—
(A) Less than 8 m (B) Upto 12 m
(C) 12 to 20 m (D) More than 20 m

94. Deep cast base beings are suitable for span—
(A) Less than 8 m (B) Upto 12 m
(C) 12 to 20 m (D) More than 20 m

95. Rocker bearings are suitable for span—
(A) Less than 8 m (B) Upto 12 m
(C) 12 to 20 m (D) More than 20 m

96. Sliding plate bearings are suitable for spans—
(A) Less than 8 m (B) Upto 12 m
(C) 12 to 20 m (D) More than 20 m

97. Trussed bridges are usually constructed as—
(A) Deck bridges
(B) Through bridges
(C) Semi-through bridges
(D) Any type of bridge

98. Steel girder bridges are mainly used on railway bridges and their maximum span should not exceeds—
(A) 10 m (B) 15 m
(C) 20 m (D) 30 m

99. Special maintenance of bridges is carried out once in—
(A) 3 to 6 months (B) 1 to 2 years
(C) 2 to 4 years (D) 3 to 5 years

100. Useful life of permanent bridge is considered to be—
(A) 15 years (B) 25 years
(C) 50 years (D) 100 years

ANSWERS

1. (C)	2. (D)	3. (B)	4. (C)	5. (A)
6. (D)	7. (D)	8. (A)	9. (A)	10. (C)
11. (D)	12. (B)	13. (B)	14. (A)	15. (D)
16. (B)	17. (D)	18. (B)	19. (D)	20. (A)
21. (D)	22. (C)	23. (D)	24. (C)	25. (B)
26. (D)	27. (D)	28. (B)	29. (C)	30. (B)
31. (D)	32. (D)	33. (A)	34. (B)	35. (A)
36. (C)	37. (C)	38. (B)	39. (C)	40. (B)
41. (D)	42. (B)	43. (D)	44. (D)	45. (D)
46. (C)	47. (B)	48. (B)	49. (D)	50. (C)
51. (C)	52. (C)	53. (B)	54. (D)	55. (C)
56. (C)	57. (C)	58. (B)	59. (C)	60. (B)
61. (C)	62. (B)	63. (C)	64. (D)	65. (A)
66. (B)	67. (D)	68. (C)	69. (A)	70. (A)
71. (D)	72. (B)	73. (A)	74. (B)	75. (D)
76. (A)	77. (D)	78. (B)	79. (C)	80. (A)
81. (C)	82. (C)	83. (C)	84. (B)	85. (C)
86. (C)	87. (A)	88. (A)	89. (B)	90. (A)
91. (D)	92. (D)	93. (B)	94. (C)	95. (D)
96. (C)	97. (B)	98. (B)	99. (D)	100. (C)

Chapter-11

ESTIMATION AND ITS EVALUATION

Estimation

An estimation is a record of computation of the quantities of materials and items of work required and expenditure likely to be incurred in the construction of building, road, dam etc.

Objectives of Estimating

(1) To find probable cost of the work.
(2) To find the quantity of various materials and labour required.
(3) To have an idea about the time of completion of the work.
(4) To help in controlling of the expenditure during execution.
(5) To help checking the work done by contractors.
(6) To help in completion of departmental formalities such as inviting tenders, technical sanction etc.
(7) To help in the fixation of standard rent, sale price of flats and valuation of properties.

Types of Estimate

Various types of estimates are :

(1) **Rough cost estimate**—It is prepared to decide the financial policy matter. It is prepared on the basis of practical knowledge and cost of similar works. The competent sanctioning authority accords 'Administrative approval'.

(2) **Plinth area estimate**—It is prepared on the basis of plinth area of a building multiplied by plinth area rate prevalent in the region. Plinth area rates are fixed from the cost of similar buildings constructed in the locality having similar finishings and amenities.

(3) **Cubical content estimate**—This estimate is worked out on the basis of the cubical contents of the proposed building to be constructed and then applying to it the rate per cubic metre. This is more accurate than plinth area estimate. The cubic content rates are deduced from the cost of similar buildings constructed in the same locality.

(4) **Detailed or Item rate estimate**—This estimate is an accurate one and is based on the plan and sanctions of the building. The quantity of items under each sub-head of work are calculated from the dimensions taken from drawing and then total cost is worked out in a form called abstract of cost.

(5) **Revised estimate**—It is also a detailed estimate and is prepared a fresh when the original sanctioned detailed estimate exceeds by 10% or more, either due to the rates being found insufficient or due to some other reasons.

(6) **Supplementary estimate**—This is a fresh detailed estimate of the additional work in addition to the original one and is prepared when additional work is required to supplement the original work.

(7) **Repair and maintenance estimate**—In order to keep the structure, roads etc. in proper condition annual repairs are carried out annually for which an estimate is prepared. The estimated amount should not be more than 1·5% of the capital cost of the work.

Measurement book

It is a note-book in which measured entries regarding the work done or supplies made are recorded. It is then used for the purpose of making payments to the contractor. This is one of the most important book and is maintained by the junior engineer.

Standard measurement book

It is a measurement book of permanent nature. In it detailed measurements of a building are correctly recorded on its completion so as to facilitate preparation of estimates for periodic repairs. For annual white washing, painting etc. payment to the contractor is made on the basis of this book after preparing the bills.

Check measurement book

It is maintained by the sub-divisional officer with the sole purpose of checking at least 25% of the measurements made by junior engineer.

Tools and plants (T and P)

In big project a percentage of 1% to $1\frac{1}{2}$% of the estimated cost is provided for the purchase of tools and plants which is required for the execution of the work.

Main Items of Work

Main items of work involved in construction are :

(a) Earth work
(b) R.C.C. and R. B. work
(c) Flooring :
  (i) Ground floor
  (ii) Upper floor
(d) Roof
(e) Plastering and Pointing
(f) Doors and windows
  (i) Frame work
  (ii) Levers and shutters for doors or windows
(g) Iron work
(h) White washing or Distempering
(i) Painting

Materials For Different Works

S. No.	Particulars of Items	For	Quantity
1.	Bricks (20 × 10 × 10 cm) in brick masonry with mortar	1 cu. m.	500 nos.
2.	Dry mortar for brick masonry	1 cu. m.	0·30 cu. m.
3.	Dry mortar for rubble stone masonry	1 cu. m.	0·40 cu. m.
4.	Stone for rubble stone masonry	1 cu. m.	1·25 cu. m.
5.	Dry mortar for cement concrete	1 cu. m.	1·54 cu. m.
6.	Brick ballast for lime concrete	1 cu. m.	1·00 cu. m.
7.	Dry mortar for lime concrete (lime and surkhi only)	1 cu. m.	0·40 cu. m.
8.	Brick required for R. B. lintels	1 cu. m.	4·50 nos.
9.	Dry mortar for R. B. work	1 cu. m.	0·45 cu. m.
10.	Dry mortar for cement pointing (deep variety)	100 sq. m.	0·80 cu. m.
11.	Dry mortar for 12·5 mm thick cement plaster	100 sq. m.	1·50 cu. m.
12.	Materials for cement concrete (1 : 2 : 4)	10 cu. m.	
	Ballast or Aggregate		8·80 cu. m.
	Sand		4·40 cu. m.
	Cement		2·20 cu. m.
13.	Material for 2·5 cm thick cement Concrete (1 : 2 : 4) floor	100 sq. m.	
	Stone grit		2·40 cu. m.
	Sand		1·20 cu. m.
	Cement		0·70 cu. m.
14.	Dry Distemper 1st coat	100 sq. m.	6 kg
15.	Dry Distemper 2nd coat	100 sq. m.	5 kg
16.	Lime for white washing one coat	100 sq. m.	10 kg
17.	Readymade paint for painting 1st coat	100 sq. m.	10 litre
18.	Cement for brick flooring	100 sq. m.	3·80 bags
19.	Dry mortar for half brick wall	100 sq. m.	3·20 cu. m.
20.	Bricks required for 10 cm thick wall	10 sq. m.	500 nos.
21.	Timber for 4 cm thick panelled door shutter	100 sq. m.	4·5 cu. m.
22.	Timber for 4 cm thick partly panelled and partly glazed shutter	100 sq. m.	3·00 cu. m.
23.	Bitumen or asphalt for painting on D.P.C. or roof		
	Ist Coat	10 sq. m.	15 kg
	IInd Coat	10 sq. m.	10 kg
24.	Galvanized iron sheets for roof	100 sq. m.	128 sq. m.
25.	A. C. corrugated sheet for roof	100 sq. m.	115 sq. m.

Task Work or Out-Turn

S. No.	Particulars of Items	Quantity per day per craftsman (Mason/Painter/Carpenter etc.)
1.	Brick work in lime or cement mortar	
	(a) In foundation and plinth	1·25 cu. m.
	(b) In super structure	1·00 cu. m.
2.	Brick work in mud mortar	
	(a) In foundation and plinth	1·50 cu. m.
	(b) In super structure	1·25 cu. m.
3.	Brick in cement or lime mortar in arches	0·55 cu. m.
4.	Half brick wall in partition	5·00 sq. m.
5.	Coursed rubble stone masonry in lime or cement mortar including dressing	0·80 cu. m.
6.	Random rubble stone masonry in lime or cement mortar	1·00 cu. m.
7.	Ashlar masonry in lime or cement mortar	0·40 cu. m.
8.	Lime concrete in foundation or floor	8·50 cu. m.
9.	Lime concrete in roof terracing	6·00 cu. m.
10.	Cement concrete 1 : 2 : 4	5·00 cu. m.
11.	R. B. work	1·00 cu. m.
12.	R. C. C. work	3·00 cu. m.
13.	12 mm plastering with cement or lime mortar	8·00 cu. m.
14.	Pointing with cement or lime mortar	10·00 sq. m.
15.	White washing or colour washing three coats	70·00 sq. m.
16.	White washing or colour washing one coat	200·00 sq. m.
17.	Painting or varnishing doors or windows one coat	25·00 sq. m.
18.	Distempering one coat	35·00 sq. m.
19.	2·5 cm c.c. floor laying	7·50 sq. m.
20.	Brick-on-edge in floor lime or cement mortar excluding L.C.	7·00 sq. m.
21.	Brick flat floor as above	8·00 sq. m.
22.	Timber framing sal or teak wood	0·07 cu. m.
23.	Country wood timber framing	0·15 sq. m.
24.	Door and window shutters panelled or glazed	0·15 sq. m.
25.	Battened door and window	0·80 sq. m.
26.	Single Allahabad tiling or Manglore tiling	6·00 sq. m.
27.	Double Allahabad tiling	4·00 sq. m.
28.	Breaking of brick ballast 40 mm size	0·75 cu. m.
29.	Breaking of brick ballast 25 mm size	0·55 cu. m.
30.	Breaking of stone ballast 40 mm size	0·40 cu. m.
31.	Ashlor stone dressing	0·70 cu. m.
32.	Flag stone dressing	1·50 sq. m.
33.	Earthwork in excavation in ordinary soil	3·00 cu. m.
34.	Excavation in rock	1·00 cu. m.

Specification

Specifications describe the nature and the class of work materials to be used in the work standard of workmanship desired etc. specification should be clear and there should not be any scope for ambiquity.

(a) **General specifications**—These give the nature and class of work in general terms to be used in the various parts of the work, from the foundation to the super structure.

(b) **Detailed specifications**—Detailed specifications of an item of work specify the quality and quantity of material, the proportions of mortar, the method of preparation and execution, workmanship etc. Detailed specifications are written to express the requirements clearly in a concise form avoiding repetition and ambiguity. These usually form a part of contract.

EXERCISE

1. The unit of payment for excavation in trenches for pipes, cables, etc. upto 1·5 m depth in ordinary rock/hard rock is in—
   (A) Cu. m. (B) Sq. m.
   (C) Metre (D) Tonne

2. Generally the unit of measurement for partition wall is—
   (A) Cu. m. (B) Sq. m.
   (C) Running metre (D) Numbers

3. The unit of payment for road side brick edging is in—
   (A) Cu. m. (B) Metre
   (C) Sq. m. (D) Kg

4. The unit of payment for blasting of rock including stacking is in—
   (A) Sq. m. (B) Cu. m.
   (C) % sq. m. (D) % cu. m.

5. The unit of measurement for electric wiring or electrification light, fan, plug points is—
   (A) Point (B) Metre
   (C) Number (D) No unit

6. The unit of measurement for steel works in trusses and its parts is in—
   (A) Quintal (B) Cm
   (C) Numbers (D) Kilogram

7. The unit of measurement for concrete work in R.C.C. is in—
   (A) Sq. m. (B) Cu. m.
   (C) 10 sq. m. (D) 10 cu. m.

8. The unit of payment for A.C. sheet roofing is in—
   (A) Cu. m. (B) Sq. m.
   (C) Metre (D) Quintal

9. The unit of payment for fixing of glass panes or cleaning is in—
   (A) Kg (B) Number
   (C) Sq. m (D) Cu. m.

10. The process of evaluating cost of construction of a project is called—
    (A) Estimate (B) Rough cost
    (C) Actual cost (D) Workable cost

11. While estimating for plastering, usually no deduction is made for—
    (A) Ends of beams
    (B) Small openings upto 0·5 sq. m.
    (C) End of rafters
    (D) All the above

12. In absence of detailed design the percentage of steel in concrete columns is usually taken as—
    (A) 10 to 15% (B) 5 to 10%
    (C) 1 to 5% (D) 15 to 20%

13. Which one is overhead expenditure of the following items ?
    (A) Rent and taxes
    (B) Establishment charges
    (C) Office stationary items
    (D) Workmen's compensation

14. In a detailed estimate the provision for contingencies is, usually—
    (A) 1% (B) 3 to 5%
    (C) 10% (D) 12 to 15%

15. Generally the provision for supervision is—
    (A) 1% (B) 2 to 3%
    (C) 3 to 5% (D) 5 to 10%

16. Estimate expected to be least accurate is—
(A) Supplementary estimate
(B) Plinth area estimate
(C) Detailed estimate
(D) Revised estimate

17. The useful part of liveable area of a building is also known as—
(A) Carpet area
(B) Circulation area
(C) Horizontal circulation area
(D) Plinth area

18. The ratio of cost of labour to the total cost of the building is—
(A) 1 : 10 (B) 1 : 4
(C) 1 : 1 (D) 6 : 10

19. The thickness of 25 gauge sheet is—
(A) Less than 1 mm
(B) 1 mm
(C) Between 1 mm and 1·5 mm
(D) 2 mm

20. In preparing cement concrete by volume the size of the wooden box used to measure sand aggregate is—
(A) 35 × 25 × 40 cm
(B) 30 × 30 × 30 cm
(C) 40 × 25 × 30 cm
(D) 35 × 30 × 40 cm

21. As a rule of thumb percentage of steel in A.C.C. columns is taken as—
(A) 1·0 to 5·0% (B) 5·0 to 10·0%
(C) 10·0 to 15·0% (D) 15·0 to 20·0%

22. In specification of earth work in foundation trenches, drains etc., specified lead is—
(A) 30 m (B) 1·5 m
(C) 50 m (D) 100 m

23. While mixing cement mortar by volume, the volume of a cement bag is specified as—
(A) 50 litres (B) 35 litres
(C) 0·050 cu. m. (D) 0·35 cu. m.

24. In measuring form work no deduction is made for opening upto—
(A) 1 m^2 (B) 0·5 m^2
(C) 0·4 m^2 (D) 0·01 m^2

25. In analysis of rates contractor profit is taken at the rate of—
(A) 1% (B) 5%
(C) 10% (D) 20%

26. How many mazdoors will be required for the disposal of 30 cu. m. of surplus earth within a lead on 30 m in one day ?
(A) 1 (B) 5
(C) 10 (D) 30

27. Number of bricks of size 20 cm × 10 cm × 10 cm required for 100 cubic metres of masonry wall is—
(A) 500 (B) 5000
(C) 50000 (D) 500000

28. Number of bricks that can be carried by 8 tonner truck is—
(A) 2500 (B) 3000
(C) 3500 (D) 4000

29. Volume of sand carried in a truck is approximately—
(A) 4 cu. m. (B) 6 cu. m.
(C) 8 cu. m. (D) 10 cu. m.

30. Number of bricks usually carried by a truck is—
(A) 2000 (B) 4000
(C) 4500 (D) 7500

31. The expected out turn for earth work in excavation in ordinary soil per mazdoor per day is—
(A) 1·00 cu. m (B) 2·00 cu. m.
(C) 3·00 cu. m. (D) 4·00 cu. m.

32. The expected out turn of 2·5 cm cement concrete floor per mason per day—
(A) 2·5 sq. m. (B) 5·0 sq. m.
(C) 7·5 sq. m. (D) 10 sq. m.

33. The expected out turn of 12 mm plastering with cement mortar is—
(A) 2·5 sq. m. (B) 4·0 sq. m.
(C) 6·0 sq. m. (D) 8·0 sq. m.

34. The expected out turn of cement concrete 1 : 2 : 4 per mason per day is—
(A) 1·5 m^3 (B) 2·5 m^3
(C) 3·5 m^3 (D) 5·0 m^3

35. In analysis of rate the quantity of dry mortar for 10 cubic metre brick work is taken as—
(A) 10 m^3 (B) 3 m^3
(C) 0·3 m^3 (D) 1 m^3

36. The quantity of stone required for 10 cubic metre of rubble stone masonry is—
(A) 10 cu. m. (B) 8 cu. m.
(C) 13·5 cu. m. (D) 12·5 cu. m.

37. Thickness of plastering is usually—
(A) 6 mm (B) 12 mm
(C) 25 mm (D) 40 mm

38. One cubic metre of mild steel weighs about—
(A) 1000 kg (B) 3625 kg
(C) 7850 kg (D) 12560 kg

39. For 100 cubic metre of roof surface, the area of A.C. corrugated sheets required will be—
(A) 80 m^2 (B) 100 m^2
(C) 115 m^2 (D) 128 m^2

40. For 100 square metres of plastered surface, the quantity of lime required for one coat of white washing will be—
(A) 10 kg (B) 6·5 kg
(C) 5 kg (D) 100 kg

41. In a detailed estimate the provision for contingencies is, usually—
(A) 1% (B) 3% to 5%
(C) 10% (D) 12% to 50%

42. Original cost of property minus depreciation is—
(A) Book value (B) Salvage value
(C) Reliable value (D) Obsolence value

43. The total estimated cost of a building electrification usually accounts for—
(A) 1% (B) 2%
(C) 8% (D) 15%

44. Capacity of a flushing cistern is usually—
(A) 10 litres (B) 25 litres
(C) 12 to 15 litres (D) 20 to 25 litres

45. Sludge from a septic tank should be removed in—
(A) 1 to 2 years (B) 3 to 5 years
(C) 3 to 4 years (D) Every year

46. The capacity of septic tank for 100 users will be roughly—
(A) 15 cu. m. (B) 7 to 8 cu. m.
(C) 50 cu. m. (D) 80 cu. m.

47. In a galvanised corrugated iron sheet the pitch (centre to centre distance) of corrugation is usually—
(A) 50 mm (B) 100 mm
(C) 75 mm (D) 150 mm

48. The standard width of asbestos cement corrugated sheet is—
(A) 0·9 m (B) 1·0 m
(C) 1·05 m (D) 1·25 m

49. The number of corrugations per sheet in case of Asbestos cement corrugated sheet is—
(A) 3 (B) 5
(C) 7 (D) 10

50. A semi-corrugated or Trafford sheet has a standard width of—
(A) 0·9 m (B) 1 m
(C) 1·05 m (D) 1·10 m

51. A work costing less than Rs. 20000 is termed as—
(A) Petty work (B) Minor work
(C) Major work (D) Casual work

ANSWERS

1. (C)	2. (B)	3. (C)	4. (B)	5. (A)
6. (A)	7. (B)	8. (B)	9. (B)	10. (A)
11. (D)	12. (C)	13. (D)	14. (B)	15. (D)
16. (B)	17. (A)	18. (B)	19. (A)	20. (A)
21. (A)	22. (A)	23. (B)	24. (C)	25. (C)
26. (C)	27. (C)	28. (A)	29. (A)	30. (B)
31. (C)	32. (C)	33. (D)	34. (D)	35. (B)
36. (D)	37. (B)	38. (C)	39. (C)	40. (A)
41. (B)	42. (A)	43. (C)	44. (C)	45. (A)
46. (B)	47. (C)	48. (C)	49. (C)	50. (D)
51. (A)				

Chapter-12

WATER RESOURCES AND WATER SUPPLY ENGINEERING

Sources of Water

Rainfall and snowfall are the primary sources of water. The secondary sources of water are :

(1) Rivers (2) Streams
(3) Springs (4) Lakes
(5) Reservoirs (6) Open wells
(7) Tube wells (8) Infiltration gallery

Climate

Climate is the general aggregate of the weather closely related to meteorology and geography. It is a composite of weather conditions over a long period of time truly portrayed.

Weather

Weather is the state of atmosphere at any given time and place. Weather generally defines states of sun shine, temperature, pressure, winds and moisture of short durations like a day, a week etc.

Precipitation

It is the total supply of water of all forms of falling moisture expressed as depth of liquid water on a horizontal surface.

Effective rainfall

It is the total net rainfall which is volumetrically equal to total direct surface runoff. It is also called rainfall excess.

Rainfall excess = Total rainfall – abstractions

Isohyet

A line joining points on a map where the amount of precipitation, recorded in a given period is the same is called an isohyet.

Rain shadow area

It is a region situated on the ice side of a mountain or mountain range where the rainfall is much less than on the windward side.

Unit storm

This is defined as a rain of such duration that the period of surface runoff is not appreciably less for any rain of shorter duration.

Interception

It is portion of the rainfall absorbed or intercepted by leaves, branches, trunks of dry trees and vegetation.

Transpiration

It is that portion of water which after circulation through plant structure goes to atmosphere in the form of vapour from the surfaces of leaves, branches etc.

Inter-flow

It is the part of infiltrated water which reappears in the streams on the surface of the earth.

Evaporation

Owing to heat energy of the sun, water is constantly changing its form to vapour from soil and water surfaces. This vapour goes to atmosphere as evaporation.

Water Resources Planning

Lakes, ponds, streams, previous soil mass and the water bearing rock formation are the best examples of natural water storages. In dry periods the natural storages release water to reduce the severity of the situation.

The absence of the natural storages of adequate capacities necessitates construction of some artificial storage works.

Storage works may be designed and constructed to serve single purpose or multipurpose. The various purposes for which storage works are required as—

(1) Irrigation
(2) Hydroelectric power generation
(3) Control of destructive floods
(4) Low water regulation for navigation
(5) Domestic and industrial water supply
(6) Recreation
(7) Preservation and breeding of useful aquative life

Types of Reservoirs

(1) Conservation reservoirs
(2) Flood control reservoirs
(3) Multipurpose reservoirs

Duty and Delta

(1) **Irrigation**—It is application of water to crops to meet water deficiency in crop production.

(2) **Infiltration**—It is the process in which wales enters the surface strata of the soil mass and moves downwards to join ground water reservoir.

(3) **Permeability**—It is that property of the soil which allows the water to move through the soil mass.

(4) **Wilting point**—The drooping down of plant leaves due to shortage of water is known as wilting point.

(5) **Field capacity**—It is the capacity of soil mass to hold the water as soil moisture at a specified time after watering.

(6) **Gross commanded area**—It is the area enclosed between the imaginary boundary line upto which certain irrigation channel is capable of supplying water for irrigation purposes.

(7) **Culturable commanded area (C.C.A.)**—It is the area of land on which crops can be grown satisfactorily.

(8) **Intensity of irrigation**—It is a ratio between cultivated land at a time in one crop season to culturable commanded area.

(9) **Duty**—It is a relationship between the amount of water and the area over which this amount of water is applied to the crop.

(10) **Base period**—It is the time elapsed from first watering at the time of sowing to the last watering when the crop matures.

Water Requirement of Crop

It is the total amount of water required at the field head to mature a crop without regard to the source of water. Obviously it includes the amount of water needed to meet—

(i) Losses through evaporation
(ii) Losses through transpiration
(iii) Plant through transpiration
(iv) Application losses
(v) Special needs

It however does not include transit losses. Water requirement of crops can be expressed mathematically as—

WR = CU + Application losses + Special needs

or WR = IR + ER + S

where WR = Water requirement of a crop
CU = Consumptive use of water
IR = Irrigation requirement of a crop at field head
ER = Effective rainfall
S = Soil moisture contribution

Methods of Irrigation

The irrigation water may be applied to the crops by three basic methods—

(a) **Surface irrigation**—In this method water is supplied on ground surface. There are number of ways for this purpose. They are :

(i) **Free flooding**—In this water is supplied till it fills the field.

(ii) **Border strip method**—In this field is divided into strips 10m to 20m wide and 100 m to 300 m long.

(iii) **Basin method**—In this one basin or rectangular check is constructed for each tree or plant. This method is useful for irrigating orchards.

(iv) **Furrow irrigation**—In this method only half to one fourth of the land is wetted.

(b) **Sprinkler irrigation**—Where irrigation water is costly, water is supplied in the form of spray through perforated pipes which may be either fixed or revolving.

(c) **Sub-surface irrigation**—In this water is supplied to sub-surface through pipe system. The purpose is to minimise evaporation losses and to effect maximum economy in water use.

Distribution System

The main components of a flow irrigation system are the following—

(i) Storage works

(ii) Diversion works

(iii) River training works

(iv) Distribution system

Parts of distribution works are :

(i) River training works

(ii) Weir

(iii) Regulator

(iv) Canal network

Depending upon the size of the canal and the function performed canals can be classified as below—

Type	Discharging Capacity m^3/sec
Minor	Less than 0·70
Major distributary	Less than 5·50
Branch	More than 5·50
Main canal	Whole discharge of a canal system

Effect of water-logging

The water logging affects the land in various ways. The various effects are the following—

1. Creation of anaerobic condition in the crop root-zone.
2. Growth of water loving wild plants
3. Impossibility of tillage operations
4. Accumulation of harmful salts
5. Reduction in time of maturity of crops

Canal lining

An irrigation canal is said to be lined when the bed and sides of a canal are protected by means of an impervious or fairly impervious material of sufficient strength. The necessity for lining a canal is felt when any one of the following requirements has to be met—

1. Canal lining controls seepage
2. Canal lining increases capacity of canals
3. Canal lining is an important anti-water-logging measure.
4. Canal lining increases commanded area for irrigation.
5. Canal lining increases available head for power generation.
6. Canal lining makes canal section stable.

Spillway

It is a structure provided in a dam to relieve the dam from excess storage of water either due to flood or due to any other reason.

The different types of spillways are—

(1) Solid gravity spillway

(2) Chute or trough spillway

(3) Saddle spillway

(4) Siphon spillway

(5) Side channel spillway

(6) Emergency spillway

(7) Shaft or Glory hole spillway

Sluices—These are the outlets provided in the dams to supply the water from the reservoir to the area below the dam.

The different types of Sluices are (or various types of outlets are)—

1. Belgaum type outlet
2. Head wall type outlet
3. Dharwar type outlet
4. Pipe outlet

Canal Head Works

The structures provided at the take off a canal to take care of irregularities in river flow condition are called canal head works.

The different parts of canal head works are—

1. Weir or an anicut or barrage
2. Divide wall
3. Under sluices or scouring sluices
4. Fish ladder
5. Head regulator for a canal
6. Silt exclusion device
7. River training works

Water Demand

Demand of water depends upon number of people in a town or city. For forecasting of population various methods are used. They are—

(1) Arithmetical increase method

(2) Geometrical increase method

(3) Incremental increase method

(4) Decreasing rate of growth method

(5) Logistic curve method

The water requirements for an average Indian town may be taken as follows—

(1) Domestic use	135 $I_{Pcd.}$
(2) Public use	25 $I_{Pcd.}$
(3) Industrial use	40 $I_{Pcd.}$
(4) Fire demand	15 $I_{Pcd.}$
(5) Wastage and losses	55 $I_{Pcd.}$
	270 $I_{Pcd.}$

The domestic demand is about 50% of the total consumption.

Fire Demand

The quantity of water for fire demand is determined by following formulae—

(1) National Board for fire underwriters formula—

$$Q = 4640\sqrt{P}\left(1 - 0{\cdot}01\sqrt{P}\right)$$

(2) Freeman's formula

$$Q = 1135{\cdot}5\left(\frac{P}{10} + 10\right)$$

(3) Kuichling's formula

$$Q = 3182\sqrt{P}$$

(4) Buston's formula

$$Q = 5663\sqrt{P}$$

where Q = Quantity of water (in litre/minute)

and P = Population of town (in thousands)

Effect of Population on Demand

The demand varies according to the population as given in the following tables—

Population of Town	Demand in Ipcd
1. 1 > 10 lakhs	200—220
2. 5–10 lakhs	180—200
3. 2–5 lakhs	160—180
4. 5000–2 lakhs	130—160
5. < 5000	100—130

Consumption of water as per Indian conditions

Purpose	Per head per day
Drinking	3 to 5 litres
Cooking	5 to 9 litres
House washing	14 to 23 litres
Baths	45 to 90 litres
Clothes washing	23 to 36 litres
WC's	27 to 36 litres

Water Born Diseases

(1) **Bacterial infections**—Typhoid fever and cholera. Paratyphoid and bacillary dysentery and hemorrhagic jaundice.

(2) **Leptospirosis**—This is known as hemorrhagic jaundice. Rats and dogs are their carriers.

(3) **Worm infections**—The eggs and larvae of intestinal worms may reach water courses from human and animal carries either directly or in washings from the soil.

(4) **Viral infections**—Virus of infections hapatitis, a number of half a dozen virus groups constituting over 100 different strains are known to be present in the faeces of infected persons.

Infections from Water-related Sources

Common modes of transmission of infection other than through drinking water are—

(i) Through watercess, or shellfish from or storedin sewage polluted water (typhoid, paratyphoid, bacillary, dysentery and infections hepatitis).

(ii) Through vegetables and fruits contaminated by faeces, sewage or sewage sludge (typhoid, paretyphoid, the dysenteries, parasitic worms, and infectious hepatitis).

(iii) Through exposure to soil contaminated by human dung (hook worm).

(iv) Through all manner of food contaminated by flies and other vermin that feed also on human faecel matter (typhoid, paratyphoid and infectious hepatitis).

(v) Through milk and milk products.

(vi) Through bathing and other exposure to polluted waters (heptospirosis and schistosomiasis).

Impurities in Water

Various substances present in water are in minerals, gases, organic matter and living organisms. These remain present in water as—

(a) Suspended impurity such as clay, silt, sand etc.

(b) Colloidal impurity such as oxides of iron, aluminium, clay, silica etc. and

(c) Dissolved impurities such as

(i) Positive ions H^+, NH_4^+, Mn^{++}, Ca^+

(ii) Negative ions Cl^-, NO_3^-, OH^-, $H_2BO_3^-$

(iii) Non-ions NH_3, H_2CO_3

Content of impurities in water is determined from the following tests—

(a) Physical test for turbidity, colour, temperature, taste, odour.

(b) Chemical tests for hardness, total solid content, chlorine, chloride, pH value, dissolved gases, metal contents, etc.

(c) Biological tests for bacteria, algae, protozoa.

Processes of Water Purification

(i) Screening

(ii) Plain sedimentation

(iii) Sedimentation aided with coagulation

(iv) Filtration and

(v) Disinfection

Disinfection of Water

It is the process of killing infective bacteria from the water and making it safe for use. It does not mean total destruction of living things in water which is known as sterilization. Disinfection of water may be done by any of the following processes—

1. Boiling of water
2. Exposure to ultra-violet rays
3. Treatment with iodine and bromine
4. Ozone treatment
5. Treatment with lime
6. Treatment with potassium permanganate
7. Chlorination

Storage and Distribution

Distribution system can be classified as—

1. Gravity system
2. Pumping system
3. Pumping and storage system

The layout of distribution system are—

(1) Dead end system

(2) Grid iron system

(3) Ring system and

(4) Radial system

EXERCISE

1. The removal of dissolved organic matter occurs in—
   (A) Slow and filters
   (B) Trickling filtres
   (C) Rapid and filters
   (D) Dual media filtres

2. Generally the height of stand pipes should be—
   (A) 15 m (B) 20 m
   (C) More than 15 m (D) Less than 15 m

3. Stand pipes are provided with—
   (A) Single inlet
   (B) Single outlet
   (C) Single inlet and outlet pipes
   (D) Double inlet and single outlet pipes

4. Pipes for water supply are tested for—
   (A) Pressure (B) Leakage
   (C) Dimensions (D) All the above

5. A.G.I. pipe has a coating of—
   (A) Zinc (B) Lead
   (C) Silver (D) Aluminium

6. The asbestos cement pipes are generally laid—
   (A) Horizontal (B) Vertical
   (C) In any position (D) Inclined at 60°

7. The least thickness of B class cast iron (spun) pipe is—
   (A) 7·2 mm (B) 7·9 mm
   (C) 8·3 mm (D) 8·6 mm

8. As compared to cast iron pipes, steel pipes are—
(A) Stronger (B) Heavier
(C) Costlier (D) All the above

9. The efficiency of a pump set is generally taken as—
(A) 90% (B) 80%
(C) 65 to 70% (D) 50 to 60%

10. For a centrifugal pump at zero discharge the head developed is—
(A) Maximum (B) Minimum
(C) Zero (D) None of these

11. The presence of ozone in water is indicated by—
(A) Red colour
(B) Light yellow colour
(C) Blue colour
(D) Black colour

12. The process used for the removal of dissolved carbon dioxide from the supply water is—
(A) Coagulation (B) Agitation
(C) Aeration (D) Sedimentation

13. The most common cause of acidity of water is—
(A) Nitrogen (B) Carbon dioxide
(C) Oxygen (D) Alkalies

14. Disinfection of water is done by—
(A) Filteration (B) Passing chlorine
(C) Ozone (D) Alum

15. The bacteria that can survive without oxygen is called—
(A) Faculative bacteria
(B) Anaerobic bacteria
(C) Aerobic bacteria
(D) Furobic bacteria

16. Aeration is done for the removal of—
(A) Colour (B) Turbidity
(C) Hardness (D) Bad odour

17. Alum increases—
(A) Hardness of water
(B) Carbonates in water
(C) Sulphates in water
(D) Acidity of water

18. Aeration is done by—
(A) Cascading
(B) Spray nozzles
(C) Tickling bed method
(D) Tank method

19. The coefficient of uniformity for slows and filters is—
(A) 1·35 (B) 1·55
(C) 1·75 (D) 2·05

20. The slow sand filter should be cleaned if the loss of head becomes more than—
(A) 0·75 m (B) 1·2 m
(C) 2·2 m (D) 3·5 m

21. The required depth of sand bed in slow sand filter should be—
(A) 0·6 to 0·9 m (B) 0·6 to 0·75 m
(C) 0·3 to 0·45 m (D) 1 to 2 m

22. The required depth of sand bed in rapid sand filter should be—
(A) 0·6 to 0·9 m (B) 0·6 to 0·75 m
(C) 0·3 to 0·45 m (D) 1 to 2 m

23. In rapid sand filters, the depth of the tank varies in between—
(A) 1 to 2m (B) 2·5 to 3·5 m
(C) 3 to 3·5 m (D) 0·6 to 0·9 m

24. Slow sand filter can be used for filtering water if its turbidity is—
(A) Not more than 50 ppm
(B) Not less than 50 ppm
(C) Not more than 80 ppm
(D) Of any value

25. The size of one unit for slow sand filter is—
(A) 30 m × 60 m (B) 8m × 10m
(C) 6m × 8m (D) 15m × 20m

26. The period of cleaning for slow sand filter is—
(A) 2 to 3 months (B) 24 to 48 months
(C) 24 to 48 months (D) 24 to 48 weeks

27. The period of cleaning for rapid sand filter is—
(A) 2–3 months (B) 24–48 months
(C) 24–48 hours (D) 24–48 weeks

28. The required percentage of total water available for washing the filter of slow sand filter is—
(A) 0·2 to 0·5% (B) 2 to 4%
(C) 0·2 to 4% (D) Zero%

29. Coagulants are used when turbidity of water is more than—
(A) 50 ppm (B) 60 ppm
(C) 70 ppm (D) 80 ppm

30. The best coagulant for removing the colour of water is—
(A) Alum (B) Lime
(C) Iron sulphate (D) Copper sulphate

31. Alum increases—
(A) Hardness of water
(B) Acidity of water
(C) Carbonates in water
(D) Sulphates in water

32. The dosage of coagulant depends upon—
(A) Turbidity (B) pH value
(C) Colour (D) All are correct

33. For sedimentation with coagulation, the detention period varies from—
(A) 1-2 hours (B) 2-4 hours
(C) 4-8 hours (D) 8-12 hours

34. Fine suspended particles are removed by—
(A) Filtration (B) Agitation
(C) Aeration (D) Bleaching

35. The organic impurities are converted into simple and harmless compounds due to—
(A) Mechanical straining
(B) Sedimentation
(C) Biological action
(D) Electrolytic action

36. The process of purifying water by passing it through a bed of fine granular material is called—
(A) Filtration (B) Coagulation
(C) Sedimentation (D) Screening

37. The void spaces in the filtering material act like—
(A) Drain
(B) Inlet
(C) Tiny settling basins
(D) Outlet

38. The path of a particle in sedimentation tank is—
(A) Parabolic (B) Straight
(C) Elliptical (D) Circular

39. The permissible velocity in pipe pressure conduit is—
(A) 0·6 to 0·9 m/s (B) 0·9 to 1·8 m/sec.
(C) 0·6 to 1·8 m/s (D) 6 to 9 m/sec.

40. The screens are arranged at an angle—
(A) 45° to 60° (B) 70° to 85°
(C) Vertically (D) 30° to 40°

41. The area of openings in screens is so kept that the velocity of flow through them does not exceed—
(A) 0·75 to 1 m/sec
(B) 1·5 to 3 m/sec
(C) 0·3 to 0·9 m/sec
(D) Less than 0·5 m/sec

42. The first stage in water treatment is—
(A) Sedimentation
(B) Filtration
(C) Disinfection
(D) Coagulation and mixing

43. Gravity conduits are generally in the form of—
(A) Open channel (B) Tunnels
(C) Aqueducts (D) All of these

44. Gravity conduits—
(A) Carry water under gravity
(B) Follow the hydraulic gradient line
(C) Are carried through tunnels in deep cuttings
(D) All the above

45. The open channels supported above the ground over trestles is called—
(A) Tunnel (B) Aqueduct
(C) Flume (D) All of these

46. Most commonly used section in the grade aqueduct is—
(A) Circular (B) Rectangular
(C) Parabolic (D) All of these

47. Grade aqueducts are not allowed to run—
(A) Are fourth full (B) Half
(C) Three fourth full (D) Full

48. The process of algae in water indicates that water is—
(A) Hard (B) Soft
(C) Acidic (D) Turbid

49. A chemical used to inhibit the growth of algae is—
(A) Chlorine (B) Iodine
(C) Sunlight (D) Algicides

50. The devices installed for drawing water from the sources are called—
(A) Filters (B) Intakes
(C) Aquifers (D) None of these

51. The total count of bacteria in potable water should not exceed—
(A) 10000 per c.c. (B) 1000 per c.c.
(C) 100 per c.c. (D) 10 per c.c.

52. Disease which is not a water borne disease is—
(A) Typhoid (B) Cholera
(C) Dysentery (D) Malaria

53. Disease caused by protozoal infections is—
(A) Poliomyelitis
(B) Amoebic dysentery
(C) Bacillary dysentery
(D) Malaria

54. The most harmful constituent of water is—
(A) Manganese (B) Bacteria
(C) Alkali (D) All of these

55. Organisms present in water causing infectious diseases are—
(A) Viruses (B) Protozoa
(C) Fungi (D) All of these

56. Which metal if present in water causes poisoning ?
(A) Mercury (B) Lead
(C) Arsenic (D) All of these

57. Lead poisoning occurs when the lead content in water exceeds—
(A) 0·05 ppm (B) 0·1 ppm
(C) 0·15 ppm (D) 0·5 ppm

58. BOD of safe drinking water should be—
(A) Nil (B) 10 ppm
(C) 20 ppm (D) 40 ppm

59. The standard BOD at 20 symbolic is taken for the consumption in—
(A) 2 days (B) 3 days
(C) 4 days (D) 5 days

60. The permissible amount of nitrites present in potable water is—
(A) Nil (B) 5 ppm
(C) 10 ppm (D) 20 ppm

61. The permissible limit of nitrate content in potable water is—
(A) 25 ppm (B) 45 ppm
(C) 100 ppm (D) 150 ppm

62. Natural mineral contaminant in water is—
(A) Calcium (B) Flourine
(C) Iron (D) Sodium

63. Mottling of teeth is associated with the presence of—
(A) Chlorides in water
(B) Flourides in water
(C) Calcium in water
(D) Sulphur in water

64. The phenolic compound in public water supply is limited to—
(A) 0·1 ppm (B) 0·01 ppm
(C) 0·001 ppm (D) 0·0001 ppm

65. The maximum allowable limit for flouride in drinking water is—
(A) 0·5 mg/litre (B) 1·0 mg/litre
(C) 1·5 mg/litre (D) 5·0 mg/litre

66. For public water supply threshold odour number should be—
(A) One (B) Between 1 and 3
(C) Three (D) More than 3

67. Maximum acidity will occur at pH value—
(A) 0 (B) 2
(C) 7 (D) 14

68. pH value representing a stronger acid is—
(A) 1 (B) 5
(C) 7 (D) 10

69. The permissible pH value for public supply water is between—
(A) 4·5 to 5·5 (B) 5·5 to 6·5
(C) 6·5 to 8·5 (D) 8·5 to 10·5

70. The most common cause of acidity in water is—
(A) Hydrogen (B) Oxygen
(C) Carbon dioxide (D) All of these

71. Dissolved impurities consist of—
(A) Bacteria (B) Iron
(C) Silt (D) Fungi

72. In water analysis, the test carried out is—
(A) Physical test (B) Chemical test
(C) Biological test (D) All of these

73. The turbidity of water is caused due to—
(A) Clay
(B) Silt
(C) Finely divided organic material
(D) All the above

74. Dissolved impurity in water is—
(A) Silt (B) Clay
(C) Calcium carbonate (D) Algae

75. The range of Hellige turbidimeter is—
(A) 0–2 ppm (B) 0–50 ppm
(C) 0–500 ppm (D) 0–1000 ppm

76. The range for Baylis turbidimeter is—
(A) 0–2 ppm (B) 0–50 ppm
(C) 0–500 ppm (D) 0–1000 ppm

77. Hard water may contain—
(A) Calcium carbonate
(B) Magnesium sulphate
(C) Magnesium bicarbonate
(D) Any of the above

78. Maximum consumption of water is noticed in—
(A) Paper factory
(B) Steel plant
(C) Automobile industry
(D) Foundary

79. The compensate for losses, thefts and wastage of water an allowance provided is—
(A) 5% (B) 10%
(C) 15% (D) 25%

80. In India the storage of water for fire on per capita basis is generally taken as—
(A) 2 litres (B) 3 litres
(C) 5 litres (D) 10 litres

81. Per capita consumption is highest in—
(A) Schools (B) Hospitals
(C) Offices (D) Factories

82. As per norms 45 litres of water per person per day is provided in case of—
(A) Hotels (B) Hospitals
(C) Office buildings (D) Public places

83. Most common surface water supply source is—
(A) Sea (B) River
(C) Pond (D) Lake

84. Which of the following gives underground water ?
(A) Spring (B) River
(C) Lake (D) Reservoir

85. The sub-surface water obtained under pressure is generally known as—
(A) Open well
(B) Tube well
(C) Infiltration well
(D) Artesian well

86. A pipe sunk into the ground to tap the underground water is called—
(A) Open well
(B) Tube well
(C) Artesian well
(D) Infilteration well

87. The specific retention is least in—
(A) Coarse gravel (B) Sand
(C) Clay (D) Silt

88. The location of a well is considered to be good if it is suck into—
(A) Coarse gravel (B) Silt
(C) Sand (D) Clay

89. The most widely used tube well in India is—
(A) Strainer well (B) Cavity well
(C) Slotted well (D) Perforated well

90. The total domestic consumption in a city as compared to total demand of the city is—
(A) 20% (B) 30%
(C) 60% (D) 75%

91. Which one of the following does not fit in with the rest ?
(A) Rain (B) Snow
(C) Frost (D) Sleet

92. For the growth of plants the useful soil moisture is—
(A) Rain water
(B) Gravity water
(C) Capillary water
(D) Chemically treated water

93. The movement of water in a channel is influenced by—
(A) Force of gravity
(B) Slope
(C) Friction of water with channel bed
(D) All the above

94. A hydrograph is a plot of—
(A) Precipitation against time
(B) Stream flow against time
(C) Surface run off against time
(D) Recorded run off against time

95. The instrument used for the measurement of wind speed is—
(A) Anemometer (B) Rotameter
(C) Odometer (D) Atmometer

96. Run off can be estimated by—
(A) Infiltration method
(B) Unit hydrograph
(C) Rational method
(D) Any of the above

97. Which of the following formula is used for Ghat area of Bombay ?
(A) English formula
(B) Lacey's formula
(C) Khosla's formula
(D) Parker's formula

98. In India the water year starts from—
(A) 1st January (B) 1st April
(C) 1st June (D) 1st October

99. The water equivalent of snowfall is taken as—
(A) 10% (B) 30%
(C) 60% (D) 75%

100. A plate of rainfall intensity versus time is known as—
(A) Hyetograph (B) Mass flow curve
(C) Duration curve (D) Unit hydrograph

101. A line joining places of equal rainfall depth in a given duration is known as—
(A) Isogonic line
(B) Isohyet
(C) Wet contour
(D) None of the above

102. Irrigation is needed where rainfall is—
(A) Abundant
(B) Just enough and timely
(C) Scarce
(D) None of these

103. Bhakra dam is built on—
(A) Beas river (B) Sutlej-river
(C) Ganga river (D) None of these

104. Flow irrigation is done from—
(A) Canals (B) Wells
(C) Tube wells (D) None of these

105. Over irrigation results in—
(A) Water-logging
(B) Wilting
(C) Fertility
(D) None of these

106. Art of irrigation is practised in India since—
(A) Ancient times (B) Moghul period
(C) British period (D) Independence

107. Jawaharlal Nehru lift irrigation scheme is located in—
(A) Rajasthan (B) U.P.
(C) Haryana (D) Punjab

108. Presently percentage of area irrigated to that of total cultivated area is—
(A) 80% (B) 33%
(C) 27% (D) 40%

109. Furrow irrigation is—
(A) Very common
(B) Unpopular
(C) Costly method of irrigation in India
(D) None of these

110. Best method of assessment of irrigation water is—
(A) Crop rate (B) Betterment levy
(C) Seasonal rate (D) Volumetric rate

111. In nature every soil sample contains—
(A) Air, water, solid matter
(B) Sand, silt and clay
(C) Both (A) and (B)
(D) None of these

112. Heavy soil is—
(A) Porous (B) Impermeable
(C) Water tight (D) None of these

113. River plains are made of—
(A) Alluvium (B) Red soils
(C) Black soils (D) None of these

114. Soil is called saline when pH value is—
(A) Less than 4 (B) 7
(C) More than 7 (D) None of these

115. Most fertile soil is—
(A) Alluvium (B) Black soil
(C) Laterite soil (D) Red soil

116. First watering to a crop is called—
(A) Paleo (B) Kor
(C) Flooding (D) None of these

117. Cash crop is generally grown during—
(A) Kharif season (B) Rabi season
(C) Both (A) and (B) (D) None of these

118. Duty of the outlet is—
(A) High
(B) Low as compared to gross duty
(C) Both (A) and (B)
(D) None of these

119. Delta of a crop means—
(A) Area under the crop
(B) Crop period
(C) Depth of water required by the crop
(D) None of these

120. Wheat belongs to same crop season as—
(A) Gram, mustard, potato, tobacco
(B) Rice, maize, spiked millet
(C) Both (A) and (B)
(D) None of these

121. Rabi crops pertain to—
(A) Winter season
(B) Summer season
(C) Monsoon season
(D) None of these

122. Duty of the canal water is expressed in—
(A) Cumec (B) Centimetres
(C) La per cumec (D) None of these

123. When you grow two crops in one year on the same field it is called—
(A) Cash crop (B) Mixed crop
(C) Double crop (D) None of these

124. A current metre is used to measure—
(A) Depth of flow
(B) Velocity of flow
(C) Peak flow
(D) None of these

125. Echo sounder is used to measure—
(A) Width of river (B) Velocity of flow
(C) Depth of flow (D) None of these

126. Velocity is measured by—
(A) Floats
(B) Current metre
(C) Pivot point method
(D) Both (A) and (B)

127. Popular method of river discharge measurement is—
(A) Area-velocity method
(B) Weir
(C) Venturi flume
(D) None of these

128. Common methods of depth measurement are—
(A) Sounding rod
(B) Echo-sounder
(C) Lead line
(D) (A), (B) and (C) are correct

129. A servoir is used for—
(A) Controlling floods
(B) Generating hydro power
(C) Irrigation
(D) (A), (B) and (C) are correct

130. Gallery is—
(A) A passage left in a completed dam
(B) An observation tower
(C) A part of a dam
(D) None of these

131. Silting of reservoir—
(A) Reduces efficiency of dam
(B) Reduces storage capacity
(C) Raises reservoir water level
(D) None of these

132. Mass curve of inflow is—
(A) Always a straight line
(B) Never a straight line
(C) A line parallel to time axis
(D) None of these

133. Economical height of dam is the one for which—
(A) Cost of dam is minimum
(B) Cost of dam per unit storage is minimum
(C) Storage is maximum
(D) None of these

134. Shape of gravity dam is roughly—
(A) Triangular (B) Square
(C) Trapezoidal (D) None of these

135. Storage dams are generally—
(A) Gravity dams (B) Earth dams
(C) Both (A) and (B) (D) None of these

136. Arch dams are constructed in—
(A) Narrow valleys with good abutment
(B) Narrow valleys with sound foundation
(C) Wide valleys with weak foundation
(D) None of these

137. Gravity dam is most suitable when foundation is—
(A) Weak
(B) Strong
(C) With heavy over burden
(D) None of these

138. Earthquake—
(A) Reduces effective weight of a dam
(B) Increases horizontal water pressure
(C) Causes dam failure by crushing
(D) None of these

139. Elementary profile of a gravity dam is—
(A) A trapezoidal figure
(B) A pyramid
(C) A right angled triangle
(D) None of these

140. Rock fill dam is—
(A) Non-rigid type (B) Rigid type
(C) High dam (D) None of these

141. Earth dam when over-topped by flood water—
(A) May not fail (B) Must fail
(C) Never fails (D) None of these

142. Earth dam could be constructed with only—
(A) Clay (B) Sand
(C) Rock (D) None of these

143. Crest of an emergency spillway is fixed at—
(A) Dead storage level
(B) F.R.L.
(C) M.W.L.
(D) Top of the dam

144. Permanent canals run—
(A) Perennially
(B) Intermittently
(C) Don't run at all
(D) Both (A) and (B)

145. The canal aligned along the water shed is known as—
(A) Ridge canal
(B) Contour canal
(C) Side slope canal
(D) None of these

146. Main canal takes-off from—
(A) A regulator (B) A river
(C) A reservoir (D) Both (B) and (C)

147. Inundation canals are—
(A) Perennial (B) Non-perennial
(C) Productive (D) None of these

148. Dowel on the canal bank serves the purpose of—
(A) A drainage (B) A foot path
(C) A road curb (D) None of these

149. Headworks can be constructed usefully in the following river reach—
 (A) Hilly
 (B) Sub-mountaineous
 (C) Plain
 (D) Both (B) and (C)

150. Fish ladder is a structure in the body of the weir used for—
 (A) Catching fish
 (B) Providing free passage to fish
 (C) Both (A) and (B)
 (D) None of these

151. Extreme condition of the meander is called—
 (A) Cut-off (B) Spur
 (C) Island (D) None of these

ANSWERS

1. (A) 2. (C) 3. (C) 4. (D) 5. (A)
6. (B) 7. (B) 8. (A) 9. (C) 10. (A)
11. (B) 12. (C) 13. (B) 14. (B) 15. (A)
16. (D) 17. (D) 18. (B) 19. (C) 20. (B)
21. (A) 22. (B) 23. (C) 24. (A) 25. (A)
26. (A) 27. (C) 28. (A) 29. (A) 30. (A)
31. (B) 32. (D) 33. (B) 34. (A) 35. (C)
36. (A) 37. (C) 38. (A) 39. (A) 40. (A)
41. (A) 42. (A) 43. (D) 44. (D) 45. (C)
46. (B) 47. (D) 48. (C) 49. (D) 50. (B)
51. (D) 52. (D) 53. (B) 54. (B) 55. (D)
56. (D) 57. (D) 58. (A) 59. (D) 60. (A)
61. (B) 62. (C) 63. (B) 64. (C) 65. (C)
66. (A) 67. (A) 68. (A) 69. (C) 70. (C)
71. (B) 72. (B) 73. (D) 74. (C) 75. (B)
76. (A) 77. (D) 78. (A) 79. (C) 80. (C)
81. (B) 82. (C) 83. (B) 84. (A) 85. (D)
86. (B) 87. (A) 88. (A) 89. (A) 90. (C)
91. (C) 92. (C) 93. (D) 94. (B) 95. (A)
96. (D) 97. (A) 98. (C) 99. (A) 100. (A)
101. (B) 102. (C) 103. (B) 104. (A) 105. (A)
106. (A) 107. (C) 108. (C) 109. (A) 110. (D)
111. (C) 112. (B) 113. (A) 114. (C) 115. (A)
116. (A) 117. (C) 118. (A) 119. (C) 120. (A)
121. (A) 122. (C) 123. (C) 124. (B) 125. (C)
126. (D) 127. (A) 128. (D) 129. (D) 130. (A)
131. (B) 132. (B) 133. (B) 134. (A) 135. (C)
136. (A) 137. (B) 138. (B) 139. (C) 140. (A)
141. (B) 142. (D) 143. (C) 144. (D) 145. (A)
146. (D) 147. (B) 148. (C) 149. (D) 150. (B)
151. (A)

Chapter-13

ENVIRONMENTAL ENGINEERING

Sewage

It is the liquid conveyed by a sewer. It may consist of any one or a mixture of liquid wastes.

Sanitary sewage

It is also known as a domestic sewage which originates in the sanitary convenience of a dwelling, business building, factory etc.

Industrial waste

It is a liquid waste from an industrial process such as dyeing, brewing or paper making.

Storm sewage

It is the liquid flowing in sewers during or following a period of rainfall and resulting therefrom as drainage water.

Bacteriology of sewage

Although sewage largely ($\approx$ 99 %) contains water, however, it also contains an ever-changing mixture of substances in solutions and suspension that are normally offensive in character and behaviour.

Sewer

It is a pipe or conduit generally closed but normally not flowing full, for carrying sewage.

Sanitary sewer

It is the conduit that carries sanitary sewage and excludes as far as possible storm sewage, surface sewage or water.

Storm sewer

It carries storm sewage including surface run-off, street wash etc.

Combined sewer

It is conduit that is designed to carry domestic sewage, industrial waste and storm sewage.

Sanitary works

The sanitary works may be classified into three needs—

(1) **Collection works**—This includes the house drainage and the network of sewers laid in the town to collect the waste from individual houses, public places and industries.

(2) **Treatment works**—This includes the processes for the treatment of sewage below disposal. The purpose is to minimize the pollution of atmosphere and water bodies in which sewage is to be disposed off.

(3) **Disposal works**—It includes arrangements for the disposal of water and sludge treated.

Self-cleaning velocity

The sewers are designed in such a way that the velocity of flow of water does not allow solids to get deposited in the sewer. Such a velocity is termed as self-cleaning velocity.

Nature of Particle in the Sewage	Self Cleaning Velocity (m/sec)
1. Rounded pebbles (12 to 15 mm)	0·5 – 0·6
2. Fine gravel	0·3
3. Coarse Sand	0·2
4. Angular stone	1·0
5. Fine sand and clay	0·15
6. Fine clay and silt	0·075

Diameter of the Sewer	Self Cleaning Velocity
1. 15 to 25 cm	1·00 m/sec
2. 30 to 60 cm	0·75 m/sec
3. Above 60 cm	0·60 m/sec

Non-scouring velocity

While the minimum velocity should be self-cleaning velocity, the velocity can only be increased upto a certain limit. It is so because if the velocity is increased too much the suspended solids may cause wear of the sewer.

Materials of the Sewer	Maximum Permissible Velocity (m/sec)
1. Cast iron sewer pipe	3·5 – 4·5
2. Earthen channel	0·6 – 1·2
3. Brick sewer pipe	1·5 – 2·5
4. Stone water sewer pipe	3·0 – 4·5
5. Concrete sewer pipe	2·4 – 3·0

Methods of Collection and Sewerage System

There are two methods of collection—

(1) **Dry or Conservancy system**—In this system accumulation of night soil in latrines or cesspools is periodically removed manually.

(2) **Water carriage system**—In this system the night soil gets mixed up with sufficient quantity of waste forming sewage which is transported through pipes for subsequent treatment and disposal in suitable manner.

Sewerage Systems are classified as follows—

(1) Combined system

(2) Separate system

(3) Partially separate system

Sewer Appurtenances

These sub-structures of sewerage system help in maintenance and operation of sewer system properly various items of this system are—

(1) Manholes (2) Clean-outs

(3) Gullies (4) Flush-tanks

(5) Oil-traps (6) Catch-basins

(7) Syphons (8) Regulators

Characteristics of Sewage

Sewage contains organic and inorganic solids in proportion of about 9 : 11. The inorganic solids are—

(1) Minerals

(2) Dissolved salts

(3) Sand

(4) Pebbles and debris

The organic solids which are present contain—

(1) Cellulose

(2) Fats and oils

(3) Nitrogenous items

(4) Hydrocarbons

Physical Characteristics

(1) **Colour**—Fresh sewage is gray in colour while decomposed sewage is dark black in colour.

(2) **Odour**—Fresh sewage gives soapy, oily or earthy odour while decomposed sewage contains foul smell of NH_3, H_2S etc.

(3) **Temperature**—The temperature of sewage is generally high than normal water, when sewage flows in closed conduits its temperature rises, resulting in increase of bacterial activity.

(4) **Solids**—Sewage generally contains 99·9% water and only 0·1% solids.

(5) **Turbidity**—It depends upon the quantity of solid matter present in the state of suspension.

Chemical Characteristics

The sewage contains organic compounds which may be categorised as—

(1) Compounds containing nitrogen such as urea proteins, amines and amino acids.

(2) Compounds free from nitrogen such as fats, soaps and carbohydrates.

Bio-chemical Oxygen Demand (B.O.D.)

The amount of oxygen required by the aerobic bacterias to cause the aerobic biological decomposition of putrescible (organic) matter of sewage for complete oxidation is called Bio-chemical Oxygen Demand (B.O.D.).

Decomposition of Sewage

Fresh sewage has oxygen content of about 2 to 5 mg/litre. Consequently the organic matter is acted upon by the aerobic and facultative bacteria play their role, resulting in spliting up of the complex organic compounds, setting free the oxygen content. Gases like ammonia, methane, hydrogen sulphide and carbon dioxide are given off. The process is called putrefaction and is the first stage in the decomposition of sewage.

Following equipment is used for putrefaction of sewage—

1. Septic tanks
2. Imhoft tanks
3. Sludge digestion tanks

For oxidation, the equipment used is—

1. Contact beds
2. Trickling filters
3. Aeration tanks

Treatment Methods

Impurity	Treatment Method
1. Floating and suspended matter	(a) Racks (b) Screens
2. Heavy and coarse suspended matter	(a) Grit chambers (b) Sedimentation tank (c) Detritus tank
3. Colloidal and dissolved organic matter	Biological growth method
4. Pathogenic bacteria	Disinfection

Sewage disposal

The effluent of sewage are disposed either in water course or on land. In the first case, it is called dilution and in the second case it is termed as sewage farming or broad irrigation.

Dilution factor

It is the ratio of moment of water for dilution w_d to the amount of sewage S and is expressed as—

$$d_f = \frac{w_d}{S}$$

where d_f = dilution factor

Sewage pumping plant

Major cities have sewage pumping plants located at one or more than one locations. The need of pumping arises when sewage from low-lying areas is to be disposed off. Pumps commonly used for sewage pumping are—

(1) Centrifugal pumps

(2) Reciprocating pumps

(3) Air pressure pumps

Industrial waste treatment and disposal

Various treatments to be provided for industrial waste are one of the following or combination of them. These are—

(1) Anaerobic digestion

(2) Chlorination

(3) Deionization

(4) Filtration

(5) Incineration

(6) Lagooning

(7) Spray irrigation

(8) Vacuum filtration

(9) Screening

(10) Precipitation

Pollutants and their sources—

Pollutant	Principal Sources
1. Carbon monoxide	Auto exhaust
2. Sulphur dioxide	Oil burners
3. Nitric oxide	Combustion
4. Nitrogen dioxide	Combustion
5. Ozone	Atmospheric photochemical reactions
6. Methane	Natural gas, decaying organic matter
7. Ethylene	Auto exhaust
8. Acetylene	Auto exhaust
9. Olefins	Auto exhaust
10. Ammonia	Decaying organic matter
11. Hydrogen sulphide	Decaying organic matters
12. Formaldehyde	Incomplete combustion

Some Important Points to be Remembered

1. Environment is a very broad term used to denote the existing surrounding which comprises of living and non-living constituents.

2. Biosphere is a segment of global environment and represent that region where life exists.

3. Separated land mass of an earth could be taken to be a bio-geographical region.

4. Biome represents a distinct characteristic combination of living and non-living components.

5. Habitat represents a local or micro environment.

6. Living component of environment is called biotic environment.

7. Non-living elements of the biosphere constitute physical environment.

8. Land, water and air or atmosphere are the three important abiotic components of the environment.

9. Small sized, self perpetuating environmental unit is called ecosystem.

10. Pollution is a state of deformity in an operating ecosystem.

11. Chemical pollutants, fuel related pollutants, pollutants from combustion of fuel, pollution from mining, solid waste disposal, waste from nuclear plants etc. are the factors which create impact on the environment.

12. Nitrogen, oxygen and other few gases are the composition of dry clean air.

13. Air pollutants may be grouped under three subheads namely gases, particular matter, radio-active materials.

14. Sulphur dioxide, nitrogen oxide, carbon monoxide, hydrocarbons, particulate matter are the major five pollutants which need immediate concern.

15. Combustion, heating and roasting process, industrial production, agricultural practices and food processing are the major activities of sources of air pollutants.

16. Ammonia, arsenic, boron, carbon monoxide, lead are the harmful effects of air pollutants.

17. Air sampling and quality measurement is undertaken at two levels as (i) ambient air sampling and (ii) stack sampling.

18. Grab sampling, absorption in liquids, absorption on solids, freeze-out sampling are the four methods for collecting gaseous pollutants.

19. Dust fall collector, high volume filtration, tape samples, impinger, electrostatic precipitation etc. are the methods used for collecting particular matter.

20. Waste water can be classified as—

(i) Drainage from irrigated agriculture

(ii) Domestic waste water and

(iii) Industrial effluents

21. Water pollutants can be grouped into following categories—

(i) Oxygen demanding wastes

(ii) Pathogenic bacteria

(iii) Non-degradable organic compound

(iv) Inorganic chemical

(v) Fertilizers

(vi) Radioactive substances

(vii) Thermal effluents

(viii) Oil spills

(ix) Sediments

EXERCISE

1. The shape of a sewer suitable to carry combined flow is—
   (A) Circular
   (B) Elliptical
   (C) Horse-shoe shaped
   (D) Egg shaped

2. The self cleaning velocity normally adopted for sewers is—
   (A) 0·1 m/sec (B) 0·2 m/sec
   (C) 0·4 m/sec (D) 0·85 m/sec

3. The minimum diameter of manhole opening is—.
   (A) 50 cm (B) 100 cm
   (C) 150 cm (D) 200 cm

4. The solid content of sewage is usually—
   (A) 99 % (B) 80 – 85 %
   (C) 10 – 15 % (D) 0·1 %

5. State sewage is usually—
   (A) Neutral (B) Acidic
   (C) Alkaline (D) Of pH value 7

6. Fresh sewage is usually—
   (A) Alkaline (B) Neutral
   (C) Acidic (D) Has pH value of 7

7. The underground conduit constructed for removal of liquid waste of a community is known as—
   (A) Canals (B) Tunnels
   (C) Sewer (D) Sewerage

8. The liquid waste conveyed by a sewer is known as—
(A) Sewer (B) Sewerage
(C) Sewage (D) All are correct

9. The waste water from bath rooms, kitchen etc. is called—
(A) Refuse (B) Sullage
(C) Sewage (D) Garbage

10. The colour of fresh sewage is—
(A) Blue (B) Green
(C) Pink (D) Grey

11. The colour of septic sewage is—
(A) Black (B) Green
(C) Grey (D) Blue

12. The minimum diameter of sewer pipe is—
(A) 1 cm (B) 5 cm
(C) 10 cm (D) 15 cm

13. In a combined sewer when only the industrial or sanitary sewage are flowing, then it is termed as—
(A) Storm water flow
(B) Industrial flow
(C) Ordinary flow
(D) Dry whether flow

14. The wash water from stables which contains animal wastes is called—
(A) Barn sludge (B) Animal sludge
(C) Live sludge (D) Horse sludge

15. Partly or completely treated sewage flowing out of a sewage treatment tank or plan is called—
(A) Treated sewage
(B) Sewage water
(C) Discharged sewage
(D) Effluent

16. A fresh sewage becomes stale in—
(A) 1 hour (B) 3 hours
(C) 6–8 hours (D) 24–36 hours

17. The process of removing suspended and colloidal matter from sewage is called—
(A) Purification (B) Clarification
(C) Suspension (D) Dewatering

18. Sewage that has received no purification treatment is called—
(A) Raw sewage (B) Untreated sewage
(C) Crude sewage (D) Fresh sewage

19. The minimum infilteration of water in one km. length of sewer pipe per day is—
(A) 0 litre (B) 500 litres
(C) 2800 litres (D) 140000 litres

20. In separate sewers the minimum velocity of flow should not be less than—
(A) 0·6 m/sec (B) 0·75 m/sec
(C) 1 m/sec (D) 6 m/sec

21. The circular section of a sewer is best suitable when diameter is—
(A) 0·75 m (B) 1·25 m
(C) 1·5 m (D) 3 m

22. While designing a sewerage system, the span of design period is generally taken as—
(A) 1 year (B) 5 years
(C) 10 years (D) 20 years

23. pH value of fresh sewage is expected to be—
(A) 1 (B) 3·5
(C) 7 (D) 8 to 11

24. The quantity of storm water from an area depends upon—
(A) Shape of the area
(B) Slope of the area
(C) Nature of soil
(D) All are correct

25. The ratio in between length of the sewer and velocity of flow when running full is known as—
(A) Inlet time
(B) Time of flow
(C) Time of concentration
(D) Time intensity

26. The minimum infiltration of water in one km, length of water pipe per day is—
(A) Zero litre (B) 500 litres
(C) 2800 litres (D) 140000 litres

27. A street sewer into which sewer from house connections in poured is called—
(A) Street sewer (B) Main sewer
(C) Lateral sewer (D) Head sewer

28. A sewer which receives its flow from a number of transverse sewers or outlets is called—
(A) Head sewer
(B) Sewer reservoir
(C) Connecting sewer
(D) Interceptor

29. The gas that is mainly responsible for explosion in sewers is—
(A) Ammonia (B) Methane
(C) Oxygen (D) Carbon monoxide

30. The lowest point of the interior of a sewer or drain at any cross-section is called—
(A) Bottom point
(B) Negative head point
(C) Revetral point
(D) Invert

31. Alignment of sewer lines is started from—
(A) The outfall (B) The high point
(C) The tail end (D) The last point

32. The proper gradient of sewer lines is ensured by—
(A) Compass (B) Plane - labelling
(C) Dumpy level (D) Telescope

33. Hydraulically the economical section of drains for large flow is—
(A) Circular (B) V-shaped
(C) Rectangular (D) Oval shaped

34. A man hole is generally provided at each—
(A) Bend
(B) Junction and change of sewer dia
(C) Change of gradient
(D) All the above

35. The minimum diameter of working chamber of a circular man hole should be—
(A) 0·5 m (B) 0·9 m
(C) 1·2 m (D) 2 m

36. Man hole covers are usually made of—
(A) Cement (B) Wood
(C) Cast iron (D) Steel

37. The location of flushing tank in a sewer in order to flush the sewer should be at its—
(A) Head (B) Middle
(C) Bottom (D) At any place

38. Bacteria in sewage are mostly—
(A) Anaerobic (B) Pathogenic
(C) Saprophytic (D) Parasitic

39. The cells of bacteria are usually in the range of—
(A) 0·1 to 0·2 microns
(B) 0·5 to 5 microns
(C) 5 to 10 microns
(D) 10 to 20 microns

40. Fungi—
(A) Grows in presence of sunlight
(B) Grows in absence of sunlight
(C) Is in organic mass
(D) Grows in lime

41. Tests for B.O.D. are usually made at—
(A) 0°C – 1 day (B) 10°C – 2 days
(C) 20°C – 5 days (D) 20°C – 3 days

42. During the treatment of sewage it is necessary to kill the—
(A) Bacterias (B) Algae
(C) Fungi (D) Pathogens

43. During putrefaction process the gas given off is—
(A) Carbon dioxide (B) Methane
(C) Ammonia (D) All of these

44. In the nitrogen cycle, ammonia is produced from—
(A) Carbohydrates (B) Cellulose
(C) Proteins (D) Sugar

45. Direct destruction to pathogenic bacteria is caused by—
(A) Direct exposure to sunlight
(B) Oxidation
(C) Chlorination
(D) Filtration

46. "The pathogens in river flows are killed and helps in self-purification." It is due to—
(A) Sunlight (B) Oxidation
(C) Reduction (D) Sedimentation

47. The depletion of dissolved oxygen due to B.O.D. reaction of sewage is called—
(A) Deoxygenation (B) Reaeration
(C) Aeration (D) Oxidation

48. In biological treatment there is formation of a biological film of—
(A) Algel (B) Anaerobic bacteria
(C) Aerobic bacteria (D) Protozoa

49. The maximum efficiency of B.O.D. removal is achieved in—
(A) Oxidation ditch (B) Trickling filters
(C) Aerated lagoons (D) Digestion tanks

50. The septic tank should be cleaned at regular interval but should not exceed—
(A) 5 years (B) 1 year
(C) 6 months (D) 3 years

51. Bacteria which use carbon dioxide as a source of carbon are known as—
(A) Autotrophic (B) Heterotrophic
(C) Aerobic (D) Anaerobic

52. If organic matter increases in sewage, the demand of oxygen will—
(A) Increase
(B) Decrease
(C) Remains constant
(D) None of these

53. In oxidation ponds, sewage is mainly treated by—
(A) Oxidation
(B) Reduction
(C) Putrefaction
(D) Aerobic bacteria action

54. A test used to measure the strength of waste water is called—
(A) Detention period (B) pH
(C) BOD (D) Surface setting rate

55. A detritus tank is provided in the primary treatment of sewage to remove—
(A) Suspended solids
(B) Grit
(C) Stones
(D) Oils and greases

56. The spacing of steel bars in coarse screens used for the treatment of sewage is—
(A) 10 mm (B) 20 mm
(C) 30 mm (D) 50 mm

57. Grit is composed mostly of—
(A) Grease (B) Paper
(C) Wood (D) Sand

58. Grit chambers are designed for particles with specific gravity of—
(A) 2·56 (B) 2·65
(C) 2·67 (D) 1·00

59. In a sludge digestion tank the process of sludge digestion is rapid when the pH value is in the range of—
(A) 4 – 7 (B) 10 – 14
(C) 6·8 – 7·2 (D) 7 – 8

60. The gas produced during sludge digestion is—
(A) Nitrogen
(B) Methane
(C) Carbon monoxide
(D) Carbon dioxide

61. Oil and grease can be removed by—
(A) Floatation tanks
(B) Rocks and screens
(C) Biological growth
(D) Disinfection

62. Fine suspended matter is removed by—
(A) Screening (B) Skimming
(C) Sedimentation (D) Filteration

63. The coagulant widely used for sewage treatment is—
(A) Ferric chloride (B) Lime
(C) Alum (D) Ferric sulphate

64. The allowable detention period in sedimentation tank is—
(A) 1 to 3 hours (B) 3 to 5 hours
(C) 5 to 10 hours (D) 12 to 24 hours

65. The B.O.D. is satisfied in biological treatment by—
(A) Screening (B) Sedimentation
(C) Filtration (D) Disinfection

66. The normal trickling filter removes—
(A) 50 % of B.O.D.
(B) 80–90 % of B.O.D.
(C) 95 % of B.O.D.
(D) 95–99 % of B.O.D.

67. Which of the following processes takes place in trickling filters ?
(A) Filteration (B) Oxidation
(C) Disinfection (D) Biological action

68. The impeller for sewage pump is—
(A) Shrouded on both sides
(B) Shrouded on one side only
(C) Open type
(D) Vane type

69. In distribution pipes, drain values are located—
(A) At junctions (B) At summit
(C) At lower joint (D) Any where

70. Which of the following disease is not considered as water borne ?
(A) Typhoid
(B) Jaundice
(C) Bacillary dysentry
(D) Malaria

71. A water closet is commonly made of—
(A) Stoneware
(B) Terracota
(C) Porcelain
(D) Glazed earthenware

72. Normally the capacity of flushing cisterns for water closets is—
(A) 1 to 2 litre (B) 2 to 5 litre
(C) 10 to 15 litre (D) 25 to 40 litre

73. The minimum height of a flushing cistern is—
(A) 1 m (B) 3 m
(C) 5 m (D) 7·5 m

74. From the septic tank the effluents are discharged into—
(A) Soak pit (B) Drainage
(C) Oxidation pond (D) Sewer

75. A trap provided to prevent the foul gases from going in the house is—
(A) Greavak trap (B) Anti-D trap
(C) Guiltry trap (D) Intercepting trap

76. For a boarding house the standard number of W.Cs. required is—
(A) 1 % (B) 2 %
(C) 3 % (D) 10 %

77. The minimum depth of water seal in a good trap is—
(A) 2·5 to 7·5 cm (B) 10 to 15 cm
(C) 15 to 25 cm (D) 30 to 35 cm

78. The average spacing of urinal stalls in public places is kept as—
(A) 30 cm (B) 60 cm
(C) 100 cm (D) 125 cm

79. For a cinema hall with seating capacity of 1000, the number of urinals to be provided is—
(A) 5 (B) 10
(C) 25 (D) 40

80. The height of a porcelain urinal stall bowl type, above the foot level is usually—
(A) 30 cm (B) 50 cm
(C) 75 cm (D) 100 cm

81. The capacity of flushing cisterns for urinals is generally—
(A) 1 litre (B) 5 litre
(C) 15 litre (D) 40 litre

82. Traps are used in household drainage systems to—
(A) Prevent entry of fond gases in the houses
(B) Restrict the flaw of water
(C) Provide partial vacuum
(D) Trap the solid wastes

83. The growth of algae is useful in—
(A) Sedimentation tank
(B) Slow and filter
(C) Oxidation pond
(D) Sludge digestion tank

84. Velocity of sound in air is around—
(A) 144 m/s (B) 244 m/s
(C) 344 m/s (D) 444 m/s

85. Which of the following can be used for disinfection of swimming pools ?
(A) Chlorine (B) Cyanurates
(C) Bromine (D) Any of these

86. Pesticide used in mosquito control is—
(A) Malathion (B) Dieldrin
(C) DDT (D) All of these

87. DDT is used by—
(A) Vapour treatment process
(B) Spray
(C) Painting
(D) Mixing with water

88. The effect of radiation on the human body depends on—
(A) Amount of body exposed
(B) Rate of absorption
(C) Metabolic state
(D) All of the above

89. A radiation, dose of 5000 roentgens is likely to result in—
(A) Vomiting
(B) Fever
(C) Change in blood colour
(D) Almost immediate incapacitation

90. Which of the following material is often used for radiation shieldings ?
(A) Wood (B) Wet cloth
(C) Lead (D) Hydrogen

91. Which of the following is fumigant pesticide ?
(A) Napthalene (B) Easters
(C) Aramite (D) All of these

92. The parts per million of oxygen in air is nearly—
(A) 20·9 (B) 209
(C) 209000 (D) 20900

93. Desalination is usually done for—
(A) Well water (B) Sea water
(C) River water (D) Lake water

94. pH value of drinking water should be in the range—
(A) 1 to 10 (B) 2 to 6
(C) 6·5 to 8·5 (D) 8·5 to 10

95. Micro straining is the process of reducing—
(A) pH value
(B) Turbidity
(C) Suspended solids
(D) Hardness

96. Activated carbon can be obtained from—
(A) Lignite (B) Wood
(C) Peat (D) Any of the above

97. Which of the following can be used for shredding ?
(A) Respers (B) Hoggers
(C) Crushers (D) Any of these

98. Boiling is usually done for—
(A) Casting
(B) Organic wastes
(C) Sheet metal waste
(D) Gaseous waste

99. Methane is—
(A) Odourless (B) Heavier than air
(C) Soluble in water (D) All of these

100. Main gas from land fills is—
(A) Oxygen
(B) Nitrogen
(C) Carbon monoxide
(D) Methane

ANSWERS

1. (D)	2. (D)	3. (A)	4. (D)	5. (B)
6. (A)	7. (C)	8. (C)	9. (B)	10. (D)
11. (D)	12. (D)	13. (D)	14. (A)	15. (D)
16. (C)	17. (B)	18. (C)	19. (C)	20. (A)
21. (C)	22. (D)	23. (D)	24. (D)	25. (B)
26. (C)	27. (C)	28. (D)	29. (B)	30. (D)
31. (B)	32. (C)	33. (A)	34. (D)	35. (C)
36. (C)	37. (A)	38. (C)	39. (B)	40. (A)
41. (C)	42. (D)	43. (D)	44. (B)	45. (C)
46. (A)	47. (A)	48. (C)	49. (A)	50. (D)
51. (A)	52. (A)	53. (D)	54. (C)	55. (B)
56. (D)	57. (D)	58. (B)	59. (C)	60. (D)
61. (A)	62. (C)	63. (A)	64. (A)	65. (C)
66. (B)	67. (D)	68. (C)	69. (C)	70. (D)
71. (C)	72. (C)	73. (B)	74. (A)	75. (D)
76. (D)	77. (A)	78. (B)	79. (C)	80. (C)
81. (B)	82. (A)	83. (C)	84. (C)	85. (D)
86. (D)	87. (B)	88. (D)	89. (D)	90. (A)
91. (A)	92. (C)	93. (B)	94. (C)	95. (C)
96. (D)	97. (D)	98. (C)	99. (A)	100. (D)

Chapter-14

PLANNING AND MANAGEMENT

Earth Work Equipment

Commonly used equipment for the earthwork may be classified as follows—

1. Tractor	2. Bull-dozers
3. Trailor	4. Scrapers
5. Elevating grader	6. Power shovels
7. Drag lines	8. Clamshells
9. Ditcher	10. Scoop
11. Dumpers	12. Hoes
13. Trenching machines	14. Shap foot rollers
15. Dumps.	

Tractor

It is used to pull or push other equipment

Dragline

Draglines are used to excavate earth and load it into hauling units such as trucks, tractor, pulled wagons, etc. or deposit on banks, dam etc.

Power shovels

Power shovels upto a capacity of 1·9 Cu, m, can be converted into dragline, by replacing the boom of the shovel with a crane boom and substituting the dragline bucket for the shovel dipper.

Hoes

The term Hoe is applied to an excavating machine of the power shovel group. Hoes are used to—

(i) Excavate below the natural surface of the ground on which machine rests.

(ii) Excavating trenches pits for basements etc. and

(iii) In general grading work.

Handling Equipment

The methods of materials handling depend upon the type of materials to handled, the quantities, frequencies and distances to be moved.

Conveyors—Conveyors are used where the loads are uniform, materials can move continuously, rate of movement and location are not likely to vary and cross traffic can be by passed by conveyor.

Overhead cranes

Overhead cranes and hoists are used where intermittent movements have to be taken up within a fixed area and materials are of variable size and weight.

Trucks and mobile cranes

Trucks, forklifts, mobile cranes are used where materials are picked up and moved intermittently through various routes.

Various Types of Handling Equipments

1. Fixed platform trucks
2. Elevating platform trucks
3. Mobile cranes
4. Fork lifts
5. Conveyors
   - (i) Gravity conveyors
   - (ii) Belt conveyors
   - (iii) Spiral chutes
   - (iv) Overhead chain conveyors
   - (v) Bucket conveyors
6. Elevators
7. Cranes
8. Hoists
9. Chain pulley blocks

10. Electromagnets
11. Dumpers
12. Loaders
13. Manipulators
14. Robots
15. Trolleys

Operation Research Technique (O.R.T.)

This scientific method employs mathematical logic to enable management to arrive a decision. Whether in an industrial production or civil engineering construction, selection of machines and equipments, criteria for repair or maintenance, cost analysis, present and future decisions etc. are obtained by O.R.T.

Method of operation research—

(1) Deterministic model

(i) Game theory

(ii) Linear programming

(iii) Sequencing

(iv) Transportation problems

(v) Dynamic programming

(2) Probabilistic model

(i) Waiting line or queuing theory

(ii) Monte carlo method

Network Techniques

Various techniques used in planning and management are **Bar chart (Grantt chart), critical path method (CPM)** and **Programme Evaluation and Review Technique (PERT)**.

(1) **Bar chart (Grantt chart)**—Mr. Henry L. Grantt evolved a chart for planning and controlling the production of ordnance factories, which is known as **Bar chart**. This chart is a graphic representation and consists of two co-ordinates. The horizontal cordinate represents the duration of times required for completion of the activities and vertical abscisa represents the respective jobs to be performed. The length of the bar shows the time the job or activity takes for its completion. For the activities taken concurrently the bars representing them are drawn parallel to each other. The durations of the activities are represented from the left to the right. Each bar is divided longitudinally into two portions. The top portion indicates the progress of completion of the activity whereas the bottom portion indicates the duration of the activity.

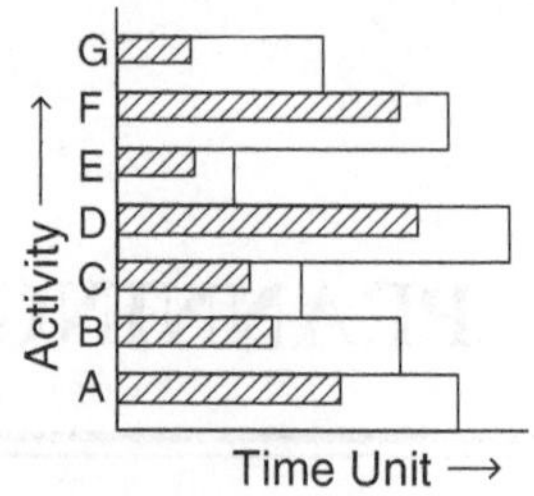

Limitations of bar chart are—

(1) Interdependencies of activities

(2) Project progress

(3) Uncertainties

(2) **Critical Path Method (CPM)**—This critical path method has been found very suitable to deal with the large and complex both civil engineering and mechanical engineering projects. This technique is useful to determine how best to reduce the time required to perform production, maintenance and construction.

The following terms are commonly used in CPM—

(i) **Network**—The diagrammatic representation of the activities of the entire project is called Network or Flow diagram.

(ii) **Activity**—An activity is a part of the project denoted by an arrow on the network. The tail of the arrow indicates the start of the activity whereas the head indicates the end of the activity.

(iii) **Event**—An event is the stage or point where all previous jobs merging in it are completed and the jobs bursting out, are still to be completed.

(iv) **Earliest Start Time (E.S.T.)**—The earliest possible time at which an activity may start is called early start time.

(v) **Earliest Finish Time (E.F.T.)**—The sum of the earliest start time of an activity and the time required for its completion is called early finish time.

(vi) **Latest Start Time (L.S.T.)**—The latest possible time at which an activity may start without delaying the date of the project is called late start time.

(vii) **Latest Finish Time (L.F.T.)**—The sum of the late start time of an activity and the time required for its completion is called late finish time.

(viii) **Total float**—The difference between the maximum time allowed for an activity and its estimated duration is called total float.

(ix) **Free float**—The duration of time by which the completion time of an activity can be delayed without affecting the start of succeeding activities is called free float.

(x) **Critical activities**—The events which have no float are called critical activities.

(xi) **Critical path**—The path in the networks, joining the critical events, is called the critical path of the work.

(3) **Programme Evaluation and Review Technique (P.E.R.T.)**—PERT also uses the project network technique, critical path and activity float. It assumes that the activities of the project and their network relationship as well defined but it allows for uncertainties in activity durations. It has following three times estimates for each activities of the network.

(i) **Optimistic time**—The estimate of the minimum possible time which an activity requires for its completion under ideal conditions is called the optimistic time.

(ii) **Most likely time**—The most likely estimate of time which an activity may take for its completion under normal conditions is called most likely time.

(iii) **Pessimistic time**—The estimate of the maximum time taken by an activity if there is delay at every stage except natural calamities is called pessimistic time.

The technical terms used in PERT analysis are—

(1) Event

(2) Activity

(3) Earliest expected time

(4) Latest allowable time

(5) Slack

(6) Critical path

(7) Variance

(8) Standard deviation

(9) Length of the project

(10) Mile stone

(11) Cost slope

EXERCISE

1. A crusher which can be used as either primary or secondary crusher is—
   (A) Hammer mill
   (B) Jaw crusher
   (C) Gyratory crusher
   (D) Cone crusher

2. Sand is produced by crushing in—
   (A) Hammer mill (B) Ball mill
   (C) Gyrator (D) Jaw crusher

3. Trenching machines can not be used for—
   (A) Rocks
   (B) Hard clay
   (C) Muddy clay
   (D) Loose material

4. When quarry rock is to be hauled, the body of a reardump truck should be—
   (A) Shallow
   (B) Deep
   (C) Shallow with sloping side boards
   (D) Deep with vertical or sloping side boards

5. Bottom dump wagons are not suitable for—
   (A) Sand
   (B) Gravels
   (C) Lumps of wet clay
   (D) Surkhi

6. Size of a dragline is indicated by the size of—
   (A) Bucket (B) Dipper
   (C) Cab (D) Boom

7. Clamshell are used to remove loose materials from inside—
   (A) Cofferdams
   (B) Pier foundations
   (C) Sewer manholes
   (D) All of these

8. Which of the following is not another name of Hoe ?
   (A) Backhoe
   (B) Back-shovel
   (C) Pull shovel
   (D) Trenching machine

9. A hand-held air-operated percussion type drill used for drilling downs holes is—
(A) Jack hammer (B) Drifter
(C) Wagon drill (D) Piston drill

10. A drill most suitable for drilling down, horizontal or up holes is—
(A) Jack hammer (B) Drifter
(C) Blasto hole drill (D) Shot drill

11. The pump suitable for high discharge pressure is—
(A) Centrifugal type (B) Screw type
(C) Rotary type (D) Reciprocating type

12. Screw pumps are used for pumping—
(A) Volatile liquids
(B) Clear water
(C) Muddy water
(D) Highly viscous liquids

13. Which information has least bearing on the selection of the pump ?
(A) Capacity in litres per minute
(B) Nature of liquid to be handled
(C) Suction lift
(D) Use of pumped fluid

14. The horse power of driving motor of a 5 cubic metre concrete mixer would be about—
(A) 1 HP (B) 5 HP
(C) 15 HP (D) 25 HP

15. Material used for the construction of load bearing piles is—
(A) Wood (B) Concrete
(C) Steel (D) Any of these

16. Which of the following equipments is not used for hauling ?
(A) Tractors (B) Bulldozers
(C) Draglines (D) Scrapers

17. Which of the following is not an excavating equipments ?
(A) Power shovel (B) Scraper
(C) Dragline (D) Hoe

18. For excavating solid rocks the most suitable equipment is—
(A) Dragline (B) Power shovel
(C) Clamshell (D) Hoe

19. The size of a power shovel is indicated by the size of—
(A) Boom (B) Hoist line
(C) Dipper (D) Cab

20. Which one of the following does not fall under the category hoisting equipment ?
(A) Chain hoist (B) Jack
(C) Hoist winch (D) Dragline

21. The construction equipment suitable for lifting and transportation is—
(A) Lift truck (B) Winch crab
(C) Spiral chute (D) Both (A) and (C)

22. Which of the following conveyor is never used ?
(A) Spiral (B) Roller
(C) Screw (D) None of these

23. A canal is trimmed on its side and bottom by—
(A) Dragline (B) Trimmer
(C) Trencher (D) Angle dozer

24. The amount saved as depreciation reserve fund will be termed as—
(A) Operating cost (B) Profit
(C) Loss (D) None of these

25. The process of removal and loading of broken rock pieces from a tunnel is known as—
(A) Mucking (B) Forepoling
(C) Jumboing (D) Bustering

26. The effectiveness of explosion during tunnelling increases by the process of—
(A) Squibbing (B) Stemming
(C) Priming (D) Firing

27. The compressor to be used for pressuring air to 10 kg is—
(A) Reciprocating type
(B) Axial flow type
(C) Centrifugal type
(D) Lobe type

28. In case of pumps the maximum suction lift is generally restricted to—
(A) 1 metre (B) 2 metre
(C) 6 metre (D) 10 metre

29. Which one of the following shovel excavator is not suitable for operation in hard soil and rocks ?
(A) Dipper shovel (B) Dragline
(C) Backhoe (D) Clamshell

30. Which one of the following shovel excavators is considered most efficient in loading carriers ?
(A) Dipper shovel
(B) Dragline
(C) Backhoe
(D) Clamshell

31. Fluid used in hydraulic braking system is—
(A) Water (B) Petrol
(C) Diesel (D) Viscous oil

32. Which one of the following is not a part of the diesel engine ?
(A) Fuel pump (B) Piston
(C) Crankshaft (D) Spark plug

33. Which one of the following is not the excavating and moving type machine ?
(A) Dragline (B) Scraper
(C) Buldozer (D) Dump truck

34. As 75 cu.m. scraper working on a 150 m haul is expected to give an output of approximately—
(A) 1 cu m/hr (B) 5 cu m/hr
(C) 10 cu m/hr (D) 100 cu m/hr

35. Which of the following is not needed in a large store ?
(A) Trolley
(B) Fork lift
(C) Gravity feed system
(D) None of these

36. The expected life of a wheeled tractor is usually—
(A) 1 year (B) 2-3 years
(C) 5-6 years (D) 8-10 years

37. The expected life of a light transport vehicle such as jeep station wagon car etc. is normally—
(A) 1000 km (B) 6000 km
(C) 16000 km (D) 20000 km

38. Which of the following is not a storage equipment ?
(A) Bins (B) Pallets
(C) Racks (D) None of these

39. 'Chute' is the name of equipment for store—
(A) Storage (B) Handling
(C) Scrap (D) Liquid transport

40. Which of the following is a material handling device ?
(A) Block and tackle (B) Skids
(C) Pipeline (D) All the above

41. The main function of radiator in a diesel engine is—
(A) To cool the engine cylinder
(B) To cool the lubricating oil
(C) To cool the fuel
(D) To cool the jacket cooling water

42. A four wheel drive has—
(A) No differential
(B) Differential on rear side only
(C) Differential on front side only
(D) Both rear as well as front differential

43. Work study comprises—
(A) Motion study
(B) Work measurement
(C) Probabilistic analysis
(D) Method study and work measurement

44. Bar charts are considered suitable for—
(A) Major projects (B) Minor projects
(C) Large projects (D) All of these

45. On a bar chart the various activities of a project are Shown by—
(A) Shaded area (B) Horizontal line
(C) Vertical line (D) Do marks

46. Jobs going ahead of schedule are conveniently shown in—
(A) Gnatt chart (B) Mile stone chart
(C) Pi chart (D) None of these

47. An event is—
(A) A definite job
(B) A definite time interval
(C) The terminals of an activity
(D) A definite position of an activity

48. Gnatt chart is drawn for—
(A) Time versus activity
(B) Activity versus activity
(C) Resources versus progress
(D) Progress versus time

49. CPM means—
(A) Computer programming mode
(B) Critical path method
(C) Controlling planning and maintenance
(D) Crucial project management

50. PERT is—
(A) Activity oriented
(B) Event oriented
(C) Time oriented
(D) Resources oriented

51. CPM network is—
(A) Event oriented
(B) Activity oriented
(C) Slack oriented
(D) Float oriented

52. Number of pre-operation associated with an operation may be—
(A) One only (B) Two only
(C) More than one (D) Three only

53. The post operations in CPM are marked—
(A) Right to an operation
(B) Left to an operation
(C) In the middle
(D) At the end

54. Latest finish time in CPM technique is indicated by a—
(A) Rectangle (B) Circle
(C) Triangle (D) Line

55. When the times of earliest finish and the latest finish are equal in an operation, the operation is called—
(A) Critical (B) Float
(C) Free float (D) All of these

56. Critical path—
(A) Is the shortest path
(B) Bad 'null' slack
(C) Had the maximum slack
(D) Had the minimum slack

57. Crashing is—
(A) Reduction is duration
(B) Reduction of resource
(C) Reduction of cost
(D) Reduction in project size

58. 'Wait' in PERT belongs to the category of—
(A) Activity (B) Node
(C) Event (D) Motion

59. In PERT technique, completion of an activity is called—
(A) Head activity (B) Head event
(C) Tail Activity (D) Tail event

60. Which of the following is a PERT event ?
(A) Transport of sand
(B) Transport of cement
(C) Concreting of roof completed
(D) Making of formwork continues

61. The technique which is activity oriented is—
(A) PERT
(B) CPM
(C) Both (A) and (B)
(D) LCES

62. Of the following the PERT event is—
(A) Fixing of doors (B) Plastering of walls
(C) Concrete cured (D) Selection of sites

63. PERT analysis is based on—
(A) Optimistic time (B) Pessimistic time
(C) Most likely time (D) All of these

64. In PERT analysis the actual performance of task is known as—
(A) Activity (B) Event
(C) Task (D) Work

65. In PERT analysis the start and end of event is represented by—
(A) Activity (B) Event
(C) Task (D) Work

66. Which of the following is an activity ?
(A) Concrete cured
(B) Walls plastered
(C) Excavation for foundation
(D) All of the above

67. The estimated time required to complete an activity is termed as—
    (A) Event (B) Duration
    (C) Active time (D) Float

68. Critical path—
    (A) Is always longest
    (B) Always shortest
    (C) May be longest
    (D) May be shortest

69. The direct cost of a project will respect to normal time is—
    (A) Minimum (B) Maximum
    (C) Zero (D) Infinite

70. Slack time refers to—
    (A) An activity
    (B) An Event
    (C) Both event and activity
    (D) None of the above

71. Indirect cost of the project—
    (A) Decrease with duration
    (B) Decrease upto optimum duration and then increase with duration
    (C) Increase with duration
    (D) All of the above

72. Total cost of the project is—
    (A) Direct cost and outage cost of project
    (B) Direct cost and overheads
    (C) Direct cost and indirect cost
    (D) None of these

73. The activity which is crashed first has—
    (A) Highest cost slope
    (B) Least cost slope
    (C) Crashing has no relation with cost slope
    (D) None of these

74. A dummy activity is that part of the project which—
    (A) Consumer resources
    (B) Doesn't consume resources
    (C) Consumes time only
    (D) None of the above

75. In PERT analysis the duration of the activities is taken as—
    (A) Optimistic time
    (B) Pessimistic time
    (C) Most likely time
    (D) Expected mean time

76. In PERT technique the critical path has slack equal to—
    (A) Zero always (B) Positive
    (C) Negative (D) Zero or negative

77. The effect of network compression is increase in the—
    (A) Total cost (B) Direct cost
    (C) Indirect cost (D) Overheads

78. The process of inspiring the subordinates to do a work or achieve an objective is called as—
    (A) Management (B) Supervision
    (C) Motivation (D) Communication

79. In a bar chart the vertical axis represents—
    (A) Time
    (B) Types of activities
    (C) Number of labours
    (D) Various activities of the project

80. The expected life of a wheeled tractor is usually—
    (A) 9 years (B) 3–5 years
    (C) 5–6 years (D) 8–10 years

81. Which one of the following bearings is not used in earth moving equipments ?
    (A) Ball bearing (B) Roller bearing
    (C) Jewel bearing (D) Bush bearing

ANSWERS

1. (A)	2. (B)	3. (A)	4. (C)	5. (C)
6. (A)	7. (D)	8. (D)	9. (A)	10. (B)
11. (D)	12. (D)	13. (D)	14. (B)	15. (D)
16. (D)	17. (B)	18. (D)	19. (C)	20. (D)
21. (D)	22. (D)	23. (B)	24. (D)	25. (A)
26. (B)	27. (A)	28. (C)	29. (D)	30. (A)
31. (D)	32. (D)	33. (D)	34. (D)	35. (D)

36. (D)	37. (C)	38. (D)	39. (B)	40. (D)
41. (D)	42. (D)	43. (D)	44. (B)	45. (B)
46. (A)	47. (D)	48. (C)	49. (B)	50. (B)
51. (B)	52. (C)	53. (A)	54. (C)	55. (A)
56. (D)	57. (A)	58. (A)	59. (B)	60. (C)
61. (B)	62. (C)	63. (D)	64. (A)	65. (B)
66. (C)	67. (B)	68. (A)	69. (A)	70. (B)
71. (C)	72. (C)	73. (B)	74. (B)	75. (D)
76. (D)	77. (B)	78. (C)	79. (D)	80. (D)
81. (C)				

Model Set-1

1. The property of the liquid surface film to exert a tension so that even small loads can be supported over it is known as—
(A) Viscosity (B) Adhesion
(C) Surface tension (D) Capillary

2. The tendency of a small drop of water to remain in a spherical form is due to the property of—
(A) Viscosity
(B) Adhesion
(C) Capillary
(D) Surface tension

3. While converting absolute unit into gravitational system 1 gram weight is equivalent to—
(A) 9810 (gm mass) (B) 981 dynes
(C) 1 dyne (D) 1000 dynes

4. To avoid correction for capillary action in manometers used for measuring pressures minimum diameter of the tube should be—
(A) 2·5 mm (B) 10 mm
(C) 50 mm (D) 6 mm

5. If 4m^3 of a certain oil weigh 3,000 kg (*f*) then specific weight of the oil will be—
(A) 750 kg/m^3 (B) 12,000 kg/m^3
(C) 7·5 gm/cm^3 (D) 75 gm/cm^3

6. The specific gravity of water at standard temperature (4° C) equal to—
(A) 0·999 (B) 1·000
(C) 1·010 (D) 0·500

7. If 4m^3 of a certain oil weighs 3000 kg. then specific gravity of the oil will be—
(A) 750 gm/cm^3 (B) 7·5
(C) 0·75 (D) 0·075

8. The attraction between molecules of a liquid and the molecules of a solid boundary surface in a contact with the liquid is known as adhesion.
(A) True (B) False

9. Intermolecular attraction between molecules of same liquid is known as cohesion.
(A) True (B) False

10. The property of a liquid by virtue of which it offers resistance to shear stress is known as surface tension.
(A) True (B) False

11. In order to avoid separation of liquid flow, absolute pressure head of water should not be less than—
(A) 1 metre (B) 9·81 metre
(C) 32 metre (D) 2·8 metre

12. In a venturimeter, the ratio between throat diameter and pipe diameter is generally adopted as—
(A) 1 : 2 (B) 1 : 4
(C) 1 : 8 (D) 2 : 11

13. The Euler's equation for steady flow of an ideal fluid along a stream line is based on Newton's—
(A) First law of forces
(B) Second law of motion
(C) Third law of momentum
(D) Law of friction

14. While applying Bernoulli equation, velocity of the liquid which is taken into account is—
(A) Initial velocity
(B) Final velocity
(C) Mean velocity
(D) Square of mean velocity

15. In a venturimeter if 'l' is the length of divergent cone and 'l_1' the length of convergent cone, then the relationship between them should be—
(A) $l = l_1$ (B) $l_1 > l$
(C) $l > 10l_1$ (D) $l = 3l_1$ to $4l_1$

16. Pitot tube is a device used in the flowing for measuring—
(A) Discharge
(B) Pressure
(C) Velocity
(D) Kinetic and potential energy

17. In a venturimeter, the approximate value of constant or Co-efficient of discharge for fluids low viscosity should be—
(A) 1 (B) 1·5
(C) 0·95 (D) 0·62

18. The formula used in pitot tube with standard abbreviations should be—
(A) $\sqrt{2\,gh}$
(B) $Ca\sqrt{2\,gh}$
(C) $\dfrac{Ca_1a_2}{\sqrt{a_1{}^2 - a_2{}^2\sqrt{2\,gh}}}$
(D) $Z_1 + \dfrac{P_1}{W} + \dfrac{V_1{}^2}{2g} = Z_2 + \dfrac{P^2}{W} + \dfrac{V_2{}^2}{2g}$

19. The force of resistance offered by an inert mass to the acceleration of a particle is known as—
(A) Friction force
(B) Pressure force
(C) Inertia force
(D) Surface tension force

20. The product of intensity of pressure and the area of a flowing liquid is known as—
(A) Friction force
(B) Pressure force
(C) Inertia force
(D) Surface tension force

21. A centrifugal pump is just the reverse of a—
(A) Radially inward flow reaction turbine
(B) Radially outward flow reaction turbine
(C) Pelton wheel
(D) Axial flow turbine

22. The direction of flow of liquid in a propeller pump impeller is in—
(A) Outward radial direction
(B) Inward radial direction
(C) Axial direction
(D) The direction tangential to the impeller

23. As compared to reciprocating pumps the discharging capacity of a centrifugal pump is more whereas its pressure head will be—
(A) Too much
(B) Same as that of reciprocating pumps
(C) Less as compared to reciprocating pump
(D) Zero

24. On a 30 cm diameter water pipe line, the operator opened the central after 15 minutes of start of the pump. As per your ideas, the result of pipe line would have been in—
(A) Bursting of the pipe line
(B) Breaking of the valve
(C) Breaking of impeller of the pump
(D) Nil *i.e.* no effect on the pipe line

25. A centrifugal pump can also be used to lifting highly viscous liquids such as oils, muddy and sewage water, paper pulps etc. whereas a reciprocating pump can lift—
(A) Pure water
(B) Muddy and sewage water
(C) Paper pulp
(D) Sugar Molasses

26. If the vanes of an impeller has only base plate and no crown plate. Then the impeller will be called as—
(A) Shrouded impeller
(B) Closed impeller
(C) Semi-open impeller
(D) Open impeller

27. When the vanes of an impeller have neither base plate, nor the crown plate, then it is known as—
(A) Shrouded impeller
(B) Closed impeller
(C) Semi-open impeller
(D) Open impeller

28. A low head pump is that one which is capable of working against a total head of—
(A) 5 m
(B) 15 m
(C) 40 m
(D) More than 40 m

29. Keeping frictional losses in mind, the vane angle in a centrifugal pump is generally kept—
(A) Not more than 20°
(B) Not less than 20°
(C) Not more than 30°
(D) Not less than 30°

30. Due to the limitations in diameter and speed of the impeller the head developed by a centrifugal pump is generally restricted to—
(A) 30 metre (B) 40 metre
(C) 50 metre (D) 100 metre

31. The water jet after striking a stationary flat plate will be defected at an angle of—
(A) 60° (B) 90°
(C) 111° (D) 135°

32. The mass of fluid issued by the jet per second for striking the fixed plate will be—
(A) WaV (B) $\frac{WaV}{g}$
(C) $WaV \cdot g$ (D) WVg

33. The force exerted by the jet of fluid on the stationary plate in the direction normal to the plate will be—
(A) $\frac{WaV^2}{g} \sin\theta$ (B) $\frac{WaV^2}{g}$
(C) $\frac{WaV}{g}$ (D) $\frac{WaV}{g} \sin\theta$

34. When water jet strikes on a number of flat plates mounted on the periphery of a wheel, then the maximum efficiency of the wheel will be—
(A) 100% (B) 80%
(C) 60% (D) 50%

35. When a fluid jet strikes on an inclined (θ) plate, the component of the thrust parallel to the direction of the jet will be—
(A) $\frac{WaV^2}{g} \sin^2\theta$
(B) $\frac{WaV^2}{g} \sin\theta \cos\theta$
(C) $\frac{WaV^2}{g} \sin\theta$
(D) $\frac{WaV^2}{g} \cos 2\theta$

36. The dynamic thrust exterted by a jet, on a flat moving plate with velocity (u) in the direction, normal to plate will be—
(A) $\frac{Wa(V-u)}{g}$ (B) $\frac{Wa(V-u)^2}{g}$
(C) $Wa(V-u)^2$ (D) $\frac{Wa(V+u)^2}{g}$

37. As compared to flat plate, force of jet on a semi-circular vane will be—
(A) Half (B) Equal
(C) Double (D) Triple

38. In a semi-circular vane, the maximum efficiency will be—
(A) 0·50 (B) 0·592
(C) 0·667 (D) 0·817

39. If the velocity of curved vane is equal to the velocity of water jet. Then the efficiency of wheel will be—
(A) Zero (B) 50%
(C) 59·20% (D) 100%

40. The maximum efficiency of semi-circular vane mounted on circular wheel will be theoretically equal to—
(A) Zero (B) 50%
(C) 59·20% (D) 100%

41. A channel is a passage through which water flows under—
(A) Force of gravity
(B) Atmospheric pressure
(C) Hydrostatic pressure
(D) Mechanical force

42. For a flow to be turbulent in an open channel Renold's number should not be less than—
(A) 500 (B) 1000
(C) 1500 (D) 2000

43. When the depth of flow changes abruptly over a short distance, then the flow will be termed as—
(A) Rapidly varied flow
(B) Unsteady flow
(C) Gradually varied flow
(D) Turbulent flow

44. Froude's number is given by the relation $Fr = \frac{V}{\sqrt{2gD}}$ and if this number is equal to 1, then the flow will be—
(A) Circular flow
(B) Subcritical flow
(C) Super critical flow
(D) Shooting flow

45. When Froude's number is more than 1, then the flow will be named as—
(A) Critical flow
(B) Subcritical flow
(C) Streaming flow
(D) Shooting flow

46. The mean velocity of flow in a channel in a vertical section will be very much nearer to the actual velocity of flow and it will be at—
(A) 0·6 depth from the bottom
(B) 0·6 depth from the top surface
(C) 0·5 depth from the top surface
(D) Free surface

47. If area of cross section of an open channel remains constant then the discharge will be maximum only when—
(A) Pressure head will be maximum
(B) Velocity of flow will be maximum
(C) Frictional loss will be minimum
(D) Bed slope will be maximum

48. A rectangular channel section will be most Economical when the depth of flow and bottom width are in ratio of—
(A) 1 : 4 (B) 1 : 1
(C) 1 : 2 (D) 2 : 1

49. For maximum mean velocity of flow in a circular channel depth of the flow should be equal to—
(A) Diameter of the channel
(B) 1·5 times the diameter of the channel
(C) 0·81 times the diameter of the channel
(D) 0·5 times the diameter of the channel

50. The thickness of boundary layer at the leading edge will be—
(A) Maximum (B) Minimum
(C) Zero (D) Average

51. The science which deals with the geometry of motion of fluids without concerning the forces causing the motion is known as—
(A) Hydro kinetic (B) Hydro kinematics
(C) Hydrostatics (D) Volumetrics

52. The flow of purified water, supplied in big cities through pipes is treated as—
(A) Steady flow
(B) Uniform flow
(C) Non-uniform flow
(D) Laminar flow

53. When various fluid particles move in layers with one layer of fluid sliding smoothly over the adjacent layer, then the flow is said to be—
(A) Laminar flow (B) Uniform flow
(C) Steady flow (D) Rotational flow

54. The line which is traced by a fluid particles passing through a fixed point in the field of flow is known as—
(A) Stream line (B) Streak line
(C) Path line (D) Vortex line

55. Water flows at the rate of 100 metre per minute in a pipe of diameter 30 cm. If this diameter of pipe is reduced to 15 cm then velocity of water in the pipe will be—
(A) 50 metre (B) 25 metre
(C) 200 metre (D) 400 metre

56. If in the case mentioned above, instead of reducing, diameter of pipe is increased to 60 cm, then the velocity of water in the pipe will be—
(A) 25 m (B) 50 m
(C) 200 m (D) 400 m

57. A line the tangent to which at any point gives the direction of motion at that point is called as—
(A) Streak line (B) Path line
(C) Stream line (D) Vortex line

58. When the fluid particles are in an extreme state of disarray their velocity fluctuation are particularly violent and erratic then the flow will be called as—
(A) Unsteady flow
(B) Turbulent flow
(C) Rotational flow
(D) All the above are correct

59. A flow in which the velocities of liquid particles at all sections of the pipe or channel are equal, is called as—
(A) Uniform flow (B) Laminar flow
(C) Turbulent flow (D) Unsteady flow

60. Bernoulli's theorem has been derived under the assumption that no external force acts on the liquids except—
(A) Atmospheric pressure
(B) Force due to gravity
(C) Wind pressure
(D) None of these

61. The knowledge of civil engineering materials is essential, to enable an engineer to—
(A) Know the source of materials used
(B) Give proper specification for purchase
(C) Make selection of suitable material for particular works
(D) All the above are correct

62. An aggregate of minerals forming the earth crust is known as—
(A) Rock (B) Stone
(C) Soil (D) Magma

63. The rock formed by the solidification of molten material either inside or on the surface of the earth will be called—
(A) Metamorphic rock
(B) Sedimentary rock
(C) Igneous rock
(D) Stratified rock

64. A natural material of construction obtained from rocks by any suitable method is called—
(A) Brick (B) Stone
(C) Timber (D) Iron

65. The rocks formed by the consolidation of particles of the pre-existing rocks under the effects of the weathering agencies are called as—
(A) Igneous rocks
(B) Sedimentary rocks
(C) Metamorphic rocks
(D) Intrusive rocks

66. The rocks which can be splitted up into thin sheets are known as—
(A) Un-stratified rocks
(B) Stratified rocks
(C) Foliated rocks
(D) Intrusive rocks

67. The rocks from the pre-existing rocks within the earth's crust under the influence of variation in temperature or pressure are called as—
(A) Igneous rocks
(B) Metamorphic rocks
(C) Intrusive rocks
(D) Eruptive rocks

68. The rocks having clay as their major constituent are known as—
(A) Argillaceous rocks
(B) Siliceous rocks
(C) Calcareous rocks
(D) Igneous rocks

69. The original position that a stone occupies when it formed is known as—
(A) The natural bed
(B) Cleavage plane
(C) Neither (A) nor (B)
(D) Both (A) and (B)

70. The colour of a mineral in its powdered form is known as—
(A) Streak (B) Lusture
(C) Texture (D) Structure

71. The appearance of a mineral in reflected light is known as—
(A) Streak (B) Lusture
(C) Texture (D) Structure

72. The coarseness of the grains of a mineral is known as—
(A) Fracture (B) Texture
(C) Structure (D) Lusture

73. The strength of building stone should not be less than—
(A) 1000 kg/cm^2 (B) 500 kg/cm^3
(C) 1500 kg/cm^2 (D) 300 kg/cm^2

74. The prosity of a good building stone should not be less than—
(A) 3% (B) 5%
(C) 10% (D) 20%

75. The hardest stone available till now is—
(A) Diamond (B) Carborundum
(C) Quartz (D) Marble

76. For finding the hardness of a material, the surface is scratched with a tool, finger, nail is used then the measurable hardness will be up—
(A) 1·5 (B) 2·5
(C) 6 (D) 0·5

77. A device containing a small amount of very sensitive and highly explosive blasting material used for initiating high explosions is called—
(A) Blast (B) Gelignite
(C) Detonator (D) Radiator

78. The hole made for blasting is first filled up with explosives and then with clay or non-inflammable as Tamping rod and this rod is made of—
(A) Steel (B) Brass
(C) Wood (D) Stainless steel

79. Blasting powder or gun-powder consists of—
(A) Potassium nitrate, sulphur, charcoal
(B) Nitroglycerine, fine sand
(C) Cotton saturated in nitric acid
(D) Potassium chlorate, nitro benzol

80. Granite Basalt and Trap are classified as—
(A) Metamorphic rock
(B) Sedimentary rock
(C) Igneous rock
(D) All the above are correct

81. The stones with crystalline and homogenous texture have weather resisting properties—
(A) Zero (B) Poor
(C) Good (D) Very good

82. Basalt and Trap are used in—
(A) Roads
(B) Dams
(C) Ornamental work
(D) Monumental works

83. The artificial material of construction in the form of clay blocks of uniform size and shape are known as—
(A) Concrete (B) Bricks
(C) Cement (D) Pozzalana cement

84. The main constituents of earth used for brick making are—
(A) Alumina
(B) Silica
(C) Silt
(D) All the above are correct

85. If iron oxide present in brick earth is more than the requirement then colour of the brick after burning will be—
(A) Dark red (B) Yellowish
(C) Dark black (D) Dark blue

86. A little quantity of lime present in brick earth in a finely divided state—
(A) Acts as flux causing sand to melt at low temperature
(B) Reduces shrinkage of bricks
(C) Helps to bind particles of bricks earth together
(D) All A, B and C are correct

87. The board used for providing frog in bricks is known as—
(A) Pallet board (B) Lamin board
(C) Fibre board (D) Stock board

88. The best temperature for burning of bricks is—
(A) 900°C (B) 1200°C
(C) 1500°C (D) 1800°C

89. As per I.S. 1077–1970, the size of ordinary brick should be—
(A) 20 × 10 × 10 cm
(B) 22·5 × 11·2 × 7·5
(C) 25 ×15 × 10 cm
(D) 19 × 9 × 9 cm

90. The approximate weight of a brick should be—
(A) 2·5 kg (B) 3·2 kg
(C) 4·5 kg (D) 5·1 kg

91. For laying one cubic metre of brick masonary the number of bricks required are—
(A) 500 (B) 1350
(C) 450 (D) 650

92. The minimum crushing strength of 'A' class brick should be—
(A) 150 kg/cm^2 (B) 105 kg/cm^2
(C) 70 kg/cm^2 (D) 50 kg/cm^2

93. The water absorption of A class brick should not be more than …… by weight—
(A) 20% (B) 22%
(C) 25% (D) 30%

94. The minimum crushing strength of second class bricks should not be less than—
    (A) 150 kg/cm^2 (B) 105 kg/cm^2
    (C) 70 kg/cm^2 (D) 50 kg/cm^2

95. The bricks made from the clay which can resist high temperature without melting or becoming soft, are known as—
    (A) Chromite bricks
    (B) Bauxite bricks
    (C) Refractory bricks
    (D) Ganister bricks

96. The first class refractory bricks should not be soften at a temperature less than—
    (A) 1300°C (B) 1700°C
    (C) 2000°C (D) 2400°C

97. The size of floor tiles generally available is—
    (A) 15 cm × 15 cm × 18 m
    (B) 20 cm × 20 cm × 20 m
    (C) 22·5 cm × 22·5 cm × 22 mm
    (D) All the above A, B, C are correct

98. The clay product, which is manufactured from refractory clay mixed with crushed pottery, powdered stone etc. by burning at high temperature, are known as—
    (A) Stoneware
    (B) Earthenware
    (C) Both (A) and (B) are correct
    (D) Neither (A) nor (B) is correct

99. The wooden mould used for casting bricks should be made of—
    (A) Shisham (B) Plywood
    (C) Deodar (D) Fibre board

100. Tolerance in brick mould is given to take care of …… after drying and burning of bricks.
    (A) Expansion (B) Pebbles
    (C) Shrinkage (D) Mortar

ANSWERS

1. (C)	2. (D)	3. (B)	4. (D)	5. (A)
6. (B)	7. (C)	8. (A)	9. (A)	10. (B)
11. (D)	12. (A)	13. (B)	14. (C)	15. (D)
16. (C)	17. (C)	18. (A)	19. (C)	20. (B)
21. (A)	22. (C)	23. (C)	24. (D)	25. (A)
26. (C)	27. (D)	28. (B)	29. (B)	30. (C)
31. (B)	32. (B)	33. (B)	34. (D)	35. (A)
36. (B)	37. (C)	38. (B)	39. (A)	40. (D)
41. (B)	42. (D)	43. (A)	44. (A)	45. (D)
46. (B)	47. (B)	48. (C)	49. (C)	50. (C)
51. (B)	52. (C)	53. (A)	54. (B)	55. (D)
56. (A)	57. (C)	58. (B)	59. (A)	60. (B)
61. (D)	62. (A)	63. (C)	64. (B)	65. (B)
66. (C)	67. (B)	68. (A)	69. (D)	70. (A)
71. (B)	72. (B)	73. (A)	74. (B)	75. (A)
76. (B)	77. (C)	78. (B)	79. (A)	80. (C)
81. (D)	82. (A)	83. (B)	84. (D)	85. (D)
86. (D)	87. (D)	88. (A)	89. (D)	90. (B)
91. (A)	92. (B)	93. (A)	94. (C)	95. (C)
96. (B)	97. (D)	98. (A)	99. (A)	100. (C)

Model Set-2

1. Cutbacks are used—
(A) After heating
(B) Without heating
(C) After dissolving in solvent
(D) All the above are wrong

2. Asphalt is an ...... material.
(A) Inflammable (B) Non-inflammable
(C) Both are correct (D) None of these

3. A hard and transparent or translucent material obtained by fusing and super-cooling an amorphous mixtures of various metallic is known as—
(A) Asbestos (B) Sunmica
(C) Glass (D) Polythene

4. If a glass is made of pure silica, then it will be—
(A) Insoluble in water
(B) Completely soluble in water
(C) Partially soluble in water
(D) All the above are wrong

5. Glass maker soap is added in suitable proportion to correct the colour of glass due to the presence of iron in raw materials of glass and it is chemically known as—
(A) Sodium carbonate
(B) Cadmium sulphate
(C) Manganese dioxide
(D) Lime

6. The melting temperature or ordinary glass is—
(A) 900°C to 1000°C
(B) 1100°C to 1200°C
(C) 1400°C to 1500°C
(D) 2000°C to 2400°C

7. Ordinary glass is insulated to—
(A) Heat
(B) Sound
(C) Electricity
(D) All the above are correct

8. The Co-efficient of expansion of glass is—
(A) High (B) Medium
(C) Low (D) Nil

9. The thermal conductivity of glans is—
(A) High (B) Medium
(C) Low (D) Nil

10. Plate glass is made by the rolling process of glass fabrication. Its surfaces are plain, parallel and its thickness varies from—
(A) 2 to 6 mm (B) 2 to 10 mm
(C) 6 to 25 mm (D) 10 to 50 mm

11. For window and door glass panes of maximum size 60 cm × 60 cm the thickness of sheet glass generally adopted is—
(A) 1·5 mm (B) 2·5 mm
(C) 3·5 mm (D) 5 mm

12. For layer size panes, the thickness of glass preferred is—
(A) 1·5 mm (B) 2·5 mm
(C) 3·5 mm (D) 5 mm

13. On sales counter, the glass generally used is of the type of—
(A) Sheet glass (B) Plate glass
(C) Laminated glass (D) Flint glass

14. For providing window screen of cars, buses and other automobile vehicles, the glass used is—
(A) Sheet glass (B) Plate glass
(C) Laminated glass (D) Flint glass

15. For sky lights and north light trusses, the glass used is—
(A) Sheet glass (B) Plate glass
(C) Wired glass (D) Laminated glass

16. For cut glass wares, ratio valves and optical glasses, the glass used is of the type of—
(A) Plate glass (B) Laminated glass
(C) Foam glass (D) Flint glass

17. While making a building air conditioned, the glass recommended for use is—
(A) Plate glass (B) Wired glass
(C) Laminated glass (D) Foam glass

18. Bullet resisting glass is a type of—
(A) Plate glass (B) Wired glass
(C) Laminated glass (D) Foam glass

19. While making coloured glasses, the colouring pigments are added in the glass at the stage of—
(A) Raw material (B) Melting
(C) Fusion (D) Cooling

20. The glass used for display windows of Jewellery shops should be of the type of—
(A) Sheet glass (B) Laminated glass
(C) Wired glass (D) Lead glass

21. The glass fitted in the panes of windows and ventilators of bathrooms laterines should be of—
(A) Sheet glass (B) Wired glass
(C) Ground glass (D) Coloured glass

22. The natural synthetic or semi-synthetic organic substances which are prepared out of natural or synthetic resins with or without moulding compounds are known as—
(A) Lacquer paints (B) Plastics
(C) Glass (D) Asphalt

23. Chemically plastics are compound of ...... with other elements such as hydrogen, nitrogen and oxygen.
(A) Sulphur (B) Phosphorus
(C) Carbon (D) Iron

24. In order to reduce the cost and increase the property of moulding and resistance to heat, fillers are added in plastic and generally the fillers adopted are—
(A) Asbestos (B) Mica
(C) Coir (D) Coconut shell
(E) All the above are correct

25. In order to impart decorative and colouring effect to plastics, pigments are added and generally the chemical added for this purpose is—
(A) Zinc oxide (B) Graphite
(C) Camphor (D) Phosphate

26. The chemicals added in plastic to accelerate the chemical reaction during the process of polymerisation are known as—
(A) Hardness (B) Polyments
(C) Catalysts (D) Plasticizers

27. The compounds added in plastic to improve toughness, flexibility, electrical properties and other mechanical properties are known as plasticizers. And out of the following, the non-plasticizers compound is—
(A) Camphor (B) Paraffins
(C) Phosphate (D) Graphite

28. The plastic which become hard and rigid when moulded and cannot be re-softened by heat are known as—
(A) Thermo-setting plastics
(B) Thermo-plastic plastics
(C) Both the above are correct
(D) Neither (A) nor (B) is correct

29. The specific gravity of plastic generally lies in between—
(A) 0·8 to 0·9 (B) 1 to 1·1
(C) 1·3 to 1·4 (D) 2 to 2·4

30. Cork is derived from the bark of—
(A) Shisham tree (B) Teak tree
(C) Toon tree (D) Oak tree

31. The specific gravity of cork is—
(A) 1·00 (B) 0·80
(C) 0·50 (D) 0·25

32. Out of the following properties of cork, one property is not correct and that is—
(A) It is lighter than air
(B) It is soft and flexible
(C) It is porous in structure
(D) It can be easily compressed

33. Glass wool is prepared from ...... glass as very fine fibres by dropping in it molten form through a very fine orifice and blowing with air or steam under high pressure.
(A) Silica glass (B) Lead glass
(C) Soda ash glass (D) Foam glass

34. The wood produced by blowing air through a stream of blast furnace slag is known as—
(A) Rock wool
(B) Glass wool

(C) Slag wool
(D) All the above are correct

35. A thread like product obtained from silica glass by dropping it in molten form through a very fine orifice and blowing with air steam under high pressure is known as—
(A) Glass wool
(B) Glass silk
(C) Fabric glass
(D) All the above are correct

36. Light weight concrete (Cellular concrete) is made by mixing crushed slag or grit, cement and—
(A) Zinc powder
(B) Aluminium powder
(C) Soap solution
(D) Plastic powder

37. Heat insulating materials are used for controlling the ...... of heat.
(A) Absorption (B) Reflection
(C) Transmission (D) Resistance

38. Sound insulating materials offer ...... to reflection and transmission of sound—
(A) Absorption (B) Acceleration
(C) Resistance (D) Resilience

39. A good insulator should have ...... texture—
(A) Porous (B) Fibrous
(C) Both are wrong (D) Both are correct

40. A good heat insulator should have ...... heat conductivity.
(A) Low (B) Medium
(C) High (D) Very high

41. The carbon content in wrought iron varies from—
(A) 0·8 to 1·5%
(B) 0·15 to 3%
(C) 0·10 to 25%
(D) 2 to 4·5%

42. The purest form of commercial iron is—
(A) Pig iron (B) Cast iron
(C) Wrought (D) Steel

43. Higher percentage of carbon ...... the strength of steel.
(A) Increase (B) Decreases
(C) Does not affect (D) Decreases

44. The metals in which iron forms the principal constituent are known as ...... metals.
(A) Ferrous (B) Non-ferrous
(C) Organic (D) None of these

45. The common non-ferrous metals are—
(A) Copper (B) Zinc
(C) Carbon (D) Phosphorus
(E) All are correct

46. The impure form of iron, containing carbon in between 2·0 to 4·5% is known as—
(A) Cast iron (B) Wrought iron
(C) Steel (D) Stainless steel

47. The properties of cast iron are—
(A) It can be permanently magnetised
(B) It can be easily forged and welded
(C) It can not be magnetised
(D) It can withstand shocks

48. To protect the top of a parapet wall from the action of rain water. It is provided with a special course and that course is known as—
(A) Drip course (B) String course
(C) Coping (D) Cornice

49. In a flat roofed building the roof is made water proof by coating hot bitumen and this coating is done at the rate of—
(A) 1 kg/m^2 (B) 1·67 kg/m^2
(C) 1·7 kg/m^2 (D) 2·5 kg/m^2

50. The opening provided in a wall at floor level for the purpose of light, air and communication is known as—
(A) Door
(B) Window
(C) Ventilator
(D) All the above are correct

51. The portion of the wall left on the face of the door or window is called as—
(A) Sill
(B) Jamb
(C) Reveal
(D) All the above are correct

52. The vertical member used in a door frame is called as—
(A) Post (B) Sill
(C) Handing style (D) Rail

53. The iron fitting used on either of the posts in a door frame to hold the frame firmly in the wall and also these fittings are embedded in the wall. Such type of fitting are called as—
(A) Hinge (B) Bolt
(C) Holdfast (D) None of these

54. The material, used in making of Choukhat frames are—
(A) Wood
(B) Bolt
(C) R.C.C.
(D) All the above are correct

55. The Aidth of rebate provided in a door frame is equal to the width of the shutter whereas depth of the rebate provided should be—
(A) 50 mm (B) 20 mm
(C) 12 or 13 mm (D) 5 to 8 mm

56. Generally in laterines kitchens and bathrooms the door used in from the type of—
(A) Ledged and braced door
(B) Flush door
(C) Partly glazed and partly panelled door
(D) Fully glazed door

57. In a panelled door the rail fixed in between the lock rail and top rail are called as—
(A) Top rail (B) Frieze rail
(C) Mullion (D) Lock rail

58. Generally the thickness of rails and styles adopted in a door frame is—
(A) 3 to 4 cm (B) 2 to 3 cm
(C) 5 to 8 cm (D) 8 to 10 cm

59. In a residential building the width of a door can not be less than—
(A) 1·2 m (B) 0·9 m
(C) 0·6 m (D) 0·75 m

60. As compared to the floor area of a room, the glass area in a window should not be less than—
(A) 25% (B) 20%
(C) 15% (D) 10%

61. A horizontal member fixed in a door or window frame for the separation of fan light is known as—
(A) Transom (B) Threshold
(C) Mullion (D) Sill

62. In air-conditioned building a door is required which can serve both the purposes, *i.e.*, opening and closing and for this purpose the most suitable type of door is—
(A) Swinging door
(B) Sliding door
(C) Rolling shutter door
(D) Revolving door

63. The horizontal member forming the bottom of the frame is called as—
(A) Transom (B) Sill
(C) Rail (D) Threshold

64. The glass used in the window whose one side is patterned while rolling which obscures direct vision but does not obstruct light is known as—
(A) Obscured glass
(B) Figured glass
(C) Rough cast glass
(D) All the above are correct

65. The window provided on the sloping roof of a building is called as—
(A) Dormer window
(B) Clere storey window
(C) Bay window
(D) Sky light window

66. The location of stair in a public building should be near—
(A) The entrance
(B) The centre
(C) The lavatory
(D) Ends of the building

67. The component of a stair on which a foot rests and affords to ascend or descend from one floor to another is known as—
(A) Step (B) Tread
(C) Landing (D) Riser

68. The horizontal portion of a step is called as—
(A) Tread (B) Going
(C) Stringer (D) Landing

69. The horizontal distance between two consecutive risers is known as—
(A) Tread (B) Going
(C) Stringer (D) Landing

70. The vertical distance between the two consecutive treads is called as—
(A) Going (B) Rise
(C) Riser (D) Pitch

71. The projection of tread beyond the riser is called as—
(A) Stringer (B) Noising
(C) Pitch (D) Baluster

72. The number of steps in a flight should not be less than—
(A) 12 (B) 3
(C) 2 (D) 5

73. The number of steps in a ordinary flight should not be more than—
(A) 12 (B) 10
(C) 15 (D) 20

74. The extreme support of a hand railing provided at the top and bottom of a flight is known as—
(A) Baluster (B) Newel Post
(C) Ballustrade (D) Barrister

75. The Bead room in a staircase should not be less than—
(A) 3·5 m (B) 3 m
(C) 2·10 m (D) 2 m

76. A step is called flier step when it is—
(A) An ordinary rectangular step
(B) A triangular step used to change direction
(C) Having curved rise and tread
(D) Inclined to the direction of the stair other than 90°

77. The under surface of a stair is called as—
(A) Soffit (B) Ceiling
(C) Waiste (D) All are correct

78. The addition support given to the noising in a wooden stair is known as—
(A) Soffit (B) Cannice
(C) Wall plate (D) Scotia

79. The approximate line of movement of people on a stair is known as walking line and from the centre of hand rail generally it is at a distance of—
(A) Zero cm (B) 30 cm
(C) 45 cm (D) 60 cm

80. In a stair the ratio of the vertical distance covered and the horizontal distance occupied by a flight is called as—
(A) Pitch (B) Gladient
(C) Line of nosing (D) Camber

81. The vertical intermediate support of a land railing is known as—
(A) Ballustrade (B) Barrister
(C) Baluster (D) Newel post

82. In a stair the best and most suitable position for providing winders is—
(A) At the bottom (B) At the top
(C) In the middle (D) Near the landing

83. According to I.S.I. the width of stair in a residential building can not be less than—
(A) 60 cm (B) 80 cm
(C) 100 cm (D) 120 cm

84. The height of hand railing above the tread should be generally in between—
(A) 30 cm to 50 cm
(B) 50 cm to 75 cm
(C) 75 cm to 80 cm
(D) 80 cm to 120 cm

85. When both the flights in a stair runs at right angles to each other, than that stair is classified as—
(A) Quarter turn stair
(B) Dog legged stair
(C) Open well stair
(D) Bifurcated stair

86. If landing is not to be counted than the number of trades as compared to risers in a stair are—
(A) One less (B) One more
(C) Same (D) Two less

87. The structure which divide a building into stages or storeys to give the space to live at different levels are termed as—
(A) Beam (B) Slab
(C) Terrace (D) Floor

88. Ground floor is that floor which is constructed—
(A) Just above the ground level
(B) Just above the basement floor
(C) Both (A) and (B) are correct
(D) Both (A) and (B) are wrong

89. Generally a 10 cm thick layer of sand is provided in the base and its purpose is—
(A) To keep out dampness
(B) To act as cushion
(C) To avoid forming of pot holes
(D) All the above are correct

90. Generally a layer in between the base and topping is used in some of the floors. This layer is named as under layer and the floors in which this layer is used are—
(A) Terrazzo floor
(B) Cement concrete floor
(C) Timber floor
(D) Flag stone floor

91. The narrow strips of wood or bands of plaster laid on the floor to act as guides or bringing whole of the work to a true—
(A) Battens (B) Screeds
(C) Dots (D) Dubbing

92. A floor is said to be noiseless when its top most layer—
(A) Does not produce noise if strucked by a hammer
(B) Does not produce noise when people pass over it
(C) Absorbs the noise produced inside the building
(D) All the above are correct

93. The cheapest floor which remains warm in winter and cold in summer is—
(A) Timber floor (B) Rubber floor
(C) Mud floor (D) Asphalt floor

94. Mud floors are not preferred in buildings because—
(A) They are cheap
(B) They resist the change in temperature of weather
(C) They are unhygienic due to cowdung
(D) They require cowdung wash periodically

95. Bricks floors are recommended to be used at—
(A) Workshops (B) Railway platforms
(C) Godowns (D) Verandahs

96. The stones used in flagstone flooring are rectangular or square in shape and have thickness of—
(A) 2 to 3 cm (B) 3 to 5 cm
(C) 5 to 7 cm (D) 10 to 15 cm

97. Flag stone floor is recommended to be used at—
(A) Railway platforms
(B) Courtyards
(C) Monumental buildings
(D) All are correct

98. While constructing a flag stone floor, the best method is to provide first two stones at—
(A) The centre
(B) Diagonally opposite corners
(C) Opposite longitudinal sides
(D) At any place in the floor

99. While laying cement concrete floor, the floor is divided into small panels. For better results side of the panel should not exceed in length by—
(A) 0·5 m (B) 1·0 m
(C) 1·5 m (D) 2 m

100. The thickness of terrazzo mix in Terrazzo floors should be approximately—
(A) 10 to 12 mm (B) 12 to 25 mm
(C) 25 to 40 mm (D) 10 cm

ANSWERS

1. (B)	2. (B)	3. (C)	4. (B)	5. (C)
6. (C)	7. (D)	8. (A)	9. (C)	10. (C)
11. (B)	12. (C)	13. (B)	14. (C)	15. (C)
16. (D)	17. (C)	18. (C)	19. (B)	20. (B)
21. (C)	22. (B)	23. (C)	24. (E)	25. (A)
26. (C)	27. (D)	28. (A)	29. (C)	30. (D)
31. (D)	32. (B)	33. (A)	34. (C)	35. (D)
36. (B)	37. (C)	38. (C)	39. (D)	40. (A)
41. (C)	42. (C)	43. (B)	44. (B)	45. (E)
46. (A)	47. (C)	48. (C)	49. (C)	50. (A)
51. (C)	52. (A)	53. (C)	54. (D)	55. (C)
56. (A)	57. (B)	58. (A)	59. (D)	60. (C)
61. (A)	62. (D)	63. (B)	64. (D)	65. (A)
66. (A)	67. (A)	68. (A)	69. (B)	70. (C)
71. (B)	72. (B)	73. (A)	74. (B)	75. (C)
76. (A)	77. (A)	78. (D)	79. (C)	80. (A)
81. (C)	82. (A)	83. (B)	84. (C)	85. (A)
86. (A)	87. (D)	88. (C)	89. (D)	90. (A)
91. (B)	92. (B)	93. (C)	94. (C)	95. (C)
96. (C)	97. (D)	98. (B)	99. (C)	100. (A)

Model Set-3

1. The structure having a place to live or to store, walls to enclose the space as well as to protect wind, sun and other unwanted elements and a roof to protect from rain and sun is known as—
(A) Civil Engineering
(B) Building
(C) Warehouse
(D) Foundation

2. While classifying the building, cinema hall should be classified as—
(A) Public building
(B) Commercial building
(C) Office building
(D) Anything else

3. The building of an engineering college should be classified as—
(A) Public building
(B) Commercial building
(C) Educational building
(D) Office building

4. The floor provided at ground level in a building should be termed as—
(A) Super structure (B) Sub-structure
(C) Both are correct (D) Both are wrong

5. The foundation in a building is provided not to—
(A) Increasing the bearing capacity of the soil
(B) Load the bearing surface at a uniform rate
(C) Distribute the load over a large area
(D) Increase overall stability of the structure

6. The causes of failure of foundation of a building can not be the—
(A) Unequal settlement of soil
(B) Withdrawal of sub-soil moisture
(C) Designed story height of the building
(D) Lateral escape of the supporting material

7. As per I.S. 875, the live load on the floors of residential buildings, hospital wards, private sitting room in a hostel, should be—
(A) 500 kg/m² (B) 400 kg/m²
(C) 300 kg/m² (D) 200 kg/m²

8. The load of furniture stored in a building is termed as—
(A) Dead load (B) Live load
(C) Both are correct (D) Both are wrong

9. While calculating the dead load of a building the weight of brick masonary is taken as—
(A) 1200 kg/m³ (B) 1600 kg/m³
(C) 1920 kg/m³ (D) 2300 kg/m³

10. The non-cohesive soil which give way to lateral escape when loaded but if confined they can be loaded considerably are known as spreading soils. Its examples are—
(A) Clayey soil (B) Sand
(C) Solid rocks (D) Common earth

11. While finding bearing capacity of the soil by Bearing plate method, the size of the steel plate used should be—
(A) 1 m × 1 m
(B) 60 cm × 60 cm
(C) 30 cm × 30 cm
(D) 10 cm × 10 cm

12. While finding bearing capacity by load impact method, if h = height from which weight is dropped w = Weight dropped, d = Depth of impression and A = Area of cross section of the weight, then bearing capacity will be—
(A) $\frac{h \times w}{d \times A}$ (B) $\frac{h \times d}{A \times w}$
(C) $\frac{h \times A}{w \times d}$ (D) $\frac{d \times A}{h \times w}$

13. The safe bearing capacity of hard rock should be—
(A) 110 + 1m^2
(B) Above 250 + 1 m^2
(C) Above 50 + 1 m^2
(D) 2 to 5 + 1 m^2

14. If P is the safe bearing capacity of soil and w is the total load coming on a wall, then width of the foundation should be—
(A) $w \times P$ (B) $\frac{w}{P}$
(C) $\frac{w}{P}\left(\frac{1-\sin\theta}{1+\sin\theta}\right)$ (D) $\frac{w}{P}\left(\frac{1-\sin\theta}{1+\sin\theta}\right)^2$

15. If w is the weight of soil per unit volume P = Safe bearing capacity of the soil and ϕ = Angle of repose of the soil, then depth of foundation will be equal to—
(A) $\frac{P}{w}\left(\frac{1-\sin\phi}{1+\sin\phi}\right)^2$ (B) $\frac{w}{P}\left(\frac{1-\sin\phi}{1+\sin\phi}\right)^2$
(C) $\frac{P}{w}\left(\frac{1+\sin\phi}{1-\sin\phi}\right)^2$ (D) $\frac{P}{w}\left(\frac{1-\sin^2\phi}{1+\sin^2\phi}\right)$

16. Generally for dry earth, dry clay and dry sand, the angle of repose taken is—
(A) 40° (B) 30°
(C) 25° (D) 15°

17. Generally for a column a workshop using Gantry cranes the foundation to be used should be—
(A) Pile foundation
(B) Grillage foundation
(C) Raft foundation
(D) Well foundation

18. For an ordinary double storey building the type of foundation to be used should be—
(A) Spread footing foundation
(B) Stepped foundation
(C) Raft foundation
(D) Any other type of foundation

19. While constructing a bridge upon a river, the foundation to be adopted in pillars should be—
(A) Well foundation
(B) Pile foundation
(C) Grillage foundation
(D) Raft foundation

20. For a bridge pier, in very deep water the foundation to be used should be—
(A) Well foundation
(B) Pile foundation
(C) Caisson foundation
(D) Coffer dam

21. When a brick is cut half, lengthwise then the so cut out bricks are call as—
(A) King closer
(B) Queen closer
(C) Bat
(D) Queen header

22. A horizontal layer of bricks laid in mortar or sometimes without mortar is known as—
(A) Bond
(B) Course
(C) Brick Masonry
(D) All the above are correct

23. The minimum thickness of wall in stone masonry cannot be less than—
(A) 10 cm (B) 20 cm
(C) 35 cm (D) 50 cm

24. The masonry in which quarry dressed or undressed stone are used, is known as—
(A) Random rubble masonry
(B) Rubble masonry
(C) Ashlar masonry
(D) Uncoursed Ashlar masonry

25. When the two blocks of stone have a tendency to be pulled a part in the direction of their lengths, then the type of joint to be used should be—
(A) Cramped joint
(B) Rebated or lap joint
(C) Tongued and grooved joint
(D) Tables joint

26. While laying bricks in a wall, frog of the brick must be kept ?
(A) Upward (B) Downward
(C) Sideward (D) In any position

27. In a single day the maximum height of masonry to be raised should not be more than—
(A) 0·5 m (B) 1 m
(C) 1·5 m (D) 2 m

28. The interior of hearting of a wall should be filled with—
(A) Stretchers (B) Headers
(C) Bats (D) Closers

29. If after every three or four stretchers one header is used in a brick masonry wall, then the bond so formed will be called as—
(A) English garden wall bond
(B) Flemish garden wall bond
(C) Stretcher bond
(D) Herring Bone Bond

30. The wall constructed along the slopes to protect the slopes of cutting from the adverse effect of atmosphere are called as—
(A) Retaining walls
(B) Breast walls
(C) Buttresses
(D) Counter-fort retaining wall

31. The road connecting the neighbouring countries is known as—
(A) National Highway
(B) Provincial Highway
(C) State Highway
(D) Express Highway

32. The professional head of public works department in a state is—
(A) Transport Minister
(B) Chief Engineer
(C) Superintending Engineer
(D) Executive Engineer

33. The road connecting one town with another town is designated as—
(A) National Highway
(B) State Highway
(C) District Road
(D) Village Road

34. For detailed distribution between houses, shops and factories, the best way of transportation is—
(A) Railway (B) Highway
(C) Airway (D) Waterway

35. The progress of any country can be Judged by the quality and net work of its—
(A) Railway (B) Highway
(C) Airway (D) Navigation

36. The Indian Road Congress was established in—
(A) 1934 (B) 1952
(C) 1947 (D) 1961

37. The road connecting villages with the towns or cities are classified as—
(A) Major district road
(B) State highway
(C) Minor district road
(D) Village roads

38. In C.P.W.D. circle incharge is designated as—
(A) Chief Engineer
(B) Superintending Engineer
(C) Executive Engineer
(D) None of these

39. An assistant engineer directly controls—
(A) Junior Engineer
(B) Work Assistants
(C) Skilled labour
(D) Executive Engineer

40. If a vehicle moves from its position while its wheels are locked, then it is known as—
(A) Skidding (B) Slipping
(C) Travelling (D) Moving

41. The revolution of wheels of a vehicle without any movement on the road is known as—
(A) Skidding (B) Slipping
(C) Travelling (D) Moving

42. As per IRC the camber on cement concrete roads should be—
(A) 1 in 60 to 72 (B) 1 in 45 to 60
(C) 1 in 20 to 24 (D) 1 in 12 to 16

43. The highest point on a carriage way is known as—
(A) Crown (B) Camber
(C) Super elevation (D) Gradient

44. The recommended safe coefficient of friction is—
(A) 1·5 (B) 0·15
(C) $\frac{1}{15}$ (D) 15

45. The portion of a road used by vehicular traffic is called as—
(A) Carriage way (B) Right of way
(C) Express way (D) Motor way

46. The rate of rise or fall of a road along its alignment is known as—
(A) Gradient (B) Camber
(C) Side Slope (D) Super elevation

47. The minimum land width required for National and State Highways in built-up area should be—
(A) 30·5 m (B) 24·4 m
(C) 15·4 m (D) 46 m

48. The convexity provided on the surface of a highway is known as—
(A) Camber (B) Gradient
(C) Super elevation (D) Curve

49. The widening of a road is not required when its radius will be—
(A) Less than 300 m
(B) More than 300 m
(C) Less than 460 m
(D) More than 460 m

50. As per I.R.C. the carriage way width for two lane traffic should be—
(A) 75 m (B) 10 m
(C) 12·5 m (D) 15 m

51. The maximum allowable super elevation is—
(A) 1 in 12 (B) 1 in 18
(C) 1 in 15 (D) 1 in 80

52. Parking lanes are provided on—
(A) Urban roads
(B) National Highways
(C) Major districts roads
(D) State Highway

53. As per IRC the maximum width of a vehicle should be—
(A) 2 m (B) 2·44 m
(C) 3·8 m (D) 158 m

54. As per Indian Road Congress, the slope of earth in cutting should be—
(A) 1 : 1 (B) 1 : 2
(C) 1 : 4 (D) 2 : 3

55. The width of shoulders provided on a road side varies in between 1·2 m and—
(A) 2·5 m (B) 1·8 m
(C) 6 m (D) 0·6 m

56. A gradient of 1° means a rise or fall of 1 metre over a horizontal distance of—
(A) 30 m (B) 45 m
(C) 57·3 m (D) 100 m

57. The permanent land width of minor district roads is generally taken as—
(A) 10 m (B) 15 m
(C) 18 m (D) 25 m

58. The alignment of a road should be cross the series of hills through—
(A) Ridge (B) Saddle
(C) Valley (D) None of these

59. While doing detailed survey, the instrument which should be adopted for measuring distance will be—
(A) Pedometer
(B) Steel tape
(C) Engineer's chain
(D) Tacheometer

60. The position of water streams are first noted while doing—
(A) Reconnaissance survey
(B) Preliminary survey
(C) Detailed survey
(D) Constructional survey

61. In a proposed alignment, the type of soil which will not be preferred is—
(A) Soft rock
(B) Ordinary rock
(C) Hard rock
(D) All the above are wrong

62. While doing soil survey, the depth for collecting samples of soil should lie in between—
(A) 0 m to 0·3 m (B) 0·3 m to 1 m
(C) 1·5 m to 3 m (D) 1 to 2 m

63. The center of the disc used for mandatory sign is kept above the ground level at a height of—
(A) 3·5 m (B) 2 m
(C) 2·5 m (D) 2·8 m

64. The difference in level between triangular disc and rectangular plate in cautionary sign should be—
(A) 50 cm (B) 10 cm
(C) 25 cm (D) 15 cm

65. The distance between pedestrian crossing and the stop line at a signalised intersection will be—
(A) 2 to 3 m (B) 8 to 10 m
(C) 1 to 2 m (D) 50 to 60 m

66. Mandatory signs are displayed on a disc having a diameter of—
(A) 30 cm (B) 60 cm
(C) 50 cm (D) 100 cm

67. Speed limit and dead slow mandatory signs should be located at ...... in advance on national and state highways.
(A) 60 *m* (B) 80 *m*
(C) 100 *m* (D) 120 *m*

68. The disc displaying mandatory signs should be white in colour and the symbols representing the signs should be red in colour.
(A) True (B) False

69. The product obtained after decomposition of rocks is known as—
(A) Clay (B) Silt
(C) Loam (D) Peat

70. Abrasion test is conducted to find—
(A) Hardness of aggregate
(B) Toughness of aggregate
(C) Durability of aggregate
(D) Strength of aggregate

71. The Railway provided just below the ground level are called—
(A) Railway
(B) Tube Railway
(C) Underground Railway
(D) Mono Railway

72. The railway provided underground at a depth of about 18 m or more are called—
(A) Railway
(B) Tube Railway
(C) Underground Railway
(D) Mono Railway

73. As compared to roadways maintenance cost of Railways is—
(A) More
(B) Less
(C) Same
(D) There can be no such comparison

74. The tractive resistance of the Tube in Railways as compared to Roadways will be—
(A) Low
(B) More
(C) Same
(D) There can be no such comparison

75. The first train was run on 18-4-1853 in India between Thana and—
(A) Delhi (B) Calcutta
(C) Madras (D) Bombay

76. The mechanical device which transfer chemical energy of fuel into mechanical energy in the form of motion is called—
(A) Railway (B) Locomotive
(C) Wagons (D) Rolling stock

77. The completed and finished Railway line on which wheeled vehicles are drawn by locomotive is known as—
(A) Railway (B) Permanent way
(C) Steel way (D) Rails

78. Gauge is the distance measured in plane of the rails between ...... of two parallel rail in a track.
(A) Centre lines
(B) Outer lines
(C) Inner lines
(D) Centre line of one rail to inner face of the other line

79. 50% of Indian Railways are laid in—
(A) Metre Gauge
(B) Narrow Gauge
(C) Broad Gauge
(D) Standard Gauge

80. In broad gauge the clear horizontal distance between the inner faces of two parallel rails forming the track is—
(A) 1 m (B) 0·792 m
(C) 0·6096 m (D) 1·676 m

81. In India narrow gauge is approximately …… of total railway line laid.
(A) 50% (B) 40%
(C) 25% (D) 10%

82. The feede gauge …… is commonly used for feeding raw materials to big government manufacturing concerns as well as to private factories such as steel plants, oil refineries, sugar factories etc.—
(A) 1 m (B) 0·792 m
(C) 0·6096 m (D) 1·676 m

83. The standard length for B.G. track in India is—
(A) 11·89 m (B) 11·8 m
(C) 19·2 m (D) 25·6 m

84. Generally the rail sections used in India are—
(A) Double headed
(B) Bull headed
(C) Flat footed
(D) All the above are correct

85. The bottom width of foot in a flat footed rail is—
(A) 66·7 mm (B) 70 mm
(C) 78·6 mm (D) 136·5 mm

86. Generally the weight 1 metre of the rails used in broad gauge is—
(A) 65 kg (B) 35 kg
(C) 45 kg (D) 55 kg

87. An expansion gap of …… is usually kept between the adjoining rail ends to allow for expansion of rail, due to variation in temperature.
(A) 6 to 8 mm
(B) 3 to 5 mm
(C) 1 to 3 mm
(D) 10 to 15 mm

88. The art of providing an outward slope of …… to the treads of wheels of rolling stock known as comming of wheels.
(A) 1 in 4 (B) 1 in 20
(C) 1 in 10 (D) 1 in 30

89. Creep is the …… movement of rails.
(A) Horizontal (B) Vertical
(C) Longitudinal (D) Diagonal

90. When a rail loses its weight by 5% it should be renewed and the old one can be used on unimportant tracks but the rails which have lost their weight by more than …… should not be used on any track under any circumstances.
(A) 8% (B) 10%
(C) 20% (D) 33·3%

91. In India, generally the sleeper density of sleepers per rail length used is—
(A) 12 (B) 18
(C) 24 (D) 30

92. The length of a wooden sleeper for B.G. track is—
(A) 2·70 m (B) 8·83 m
(C) 1·52 m (D) 1·676 m

93. Steel sleepers are …… shaped in section.
(A) Through (B) Rectangular
(C) Oval (D) Semi-spherical

94. The members laid transversely under the rails for supporting and fixing them at the gauge distance apart are known as—
(A) Ballast (B) Fastenings
(C) Fish plate (D) Sleepers

95. The process of connecting a track length by an electric circuit is known as track circuiting and the sleepers which are suitable for it are—
(A) Wooden sleepers
(B) Steel sleepers
(C) Cast iron sleepers
(D) All the above are suitable

96. Generally the life of wooden sleepers is known as—
(A) 2 to 3 years
(B) 5 to 8 years
(C) 12 to 15 years
(D) 35 to 50 years

97. The granular material spread on the formation of a railway track for the sleepers to rest upon is known as—
(A) Subgrade (B) Chairs
(C) Anchors (D) Ballast

98. For wooden sleepers, size of the ballast should be—
(A) 51 cm (B) 3·8 cm
(C) 2·54 cm (D) 10·16 cm

99. The depth of ballast section under the sleeper for B.C. track as per Indian standard should be—
(A) 150 mm (B) 150 to 200 mm
(C) 200 to 250 mm (D) 250 to 300 mm

100. As per Indian Standard the recommended width of ballast section at foot level of rail of BG tracks should be—
(A) 3·35 m (B) 2·25 m
(C) 1·83 m (D) 4·38 m

ANSWERS

1. (B)	2. (A)	3. (C)	4. (A)	5. (A)
6. (C)	7. (D)	8. (B)	9. (C)	10. (B)
11. (C)	12. (A)	13. (B)	14. (B)	15. (A)
16. (B)	17. (B)	18. (A)	19. (A)	20. (C)
21. (B)	22. (B)	23. (C)	24. (A)	25. (A)
26. (A)	27. (C)	28. (B)	29. (B)	30. (B)
31. (A)	32. (B)	33. (B)	34. (B)	35. (B)
36. (A)	37. (D)	38. (B)	39. (A)	40. (A)
41. (B)	42. (A)	43. (A)	44. (B)	45. (A)
46. (B)	47. (B)	48. (A)	49. (D)	50. (A)
51. (B)	52. (A)	53. (B)	54. (A)	55. (B)
56. (C)	57. (C)	58. (B)	59. (B)	60. (A)
61. (C)	62. (C)	63. (D)	64. (C)	65. (B)
66. (B)	67. (D)	68. (B)	69. (A)	70. (A)
71. (C)	72. (B)	73. (B)	74. (A)	75. (D)
76. (B)	77. (B)	78. (C)	79. (C)	80. (D)
81. (D)	82. (C)	83. (B)	84. (C)	85. (D)
86. (D)	87. (A)	88. (B)	89. (C)	90. (B)
91. (B)	92. (A)	93. (A)	94. (D)	95. (A)
96. (C)	97. (D)	98. (A)	99. (A)	100. (A)

Model Set-4

1. A mixture of natural pyrogenous hydrocarbons and their non-metallic derivative which may be gaseous, liquid viscous or solid but must be completely soluble in carbon disulphide is known as—
(A) Petroleum
(B) Bitumen
(C) Cement
(D) Cut-back emulsion

2. The different materials used for roads construction are soil, binder and—
(A) Stone aggregate (B) Sand
(C) Bitumen (D) Cement

3. In earthen roads, the common stabilizer used are—
(A) Cement
(B) Bitumen
(C) Lime
(D) All the above are correct

4. The residual product obtained by .... distillation of oil, wood, coal etc. is known as tar.
(A) Destructive
(B) Fractional
(C) Natural
(D) All the above are correct

5. ...... test is carried out to know the resistance to flow, of the given bituminous material.
(A) Softening point (B) Penetration
(C) Ductility (D) Viscosity

6. The specific gravity of bitumen is 0·99 whereas for asphalt it varies from 1·26 to 1·31.
(A) True (B) False

7. The water carried fine grained soil, constructing an intimate mixture of clay, sand etc. is known as—
(A) Clay (B) Loam
(C) Silt (D) Peat

8. The camber and gradient on earthen roads should not be more than—
(A) 1 in 10 (B) 1 in 20
(C) 1 in 50 (D) 1 in 80

9. The cross slopes on earthen roads should not be less than—
(A) 1 in 10 (B) 1 in 20
(C) 1 in 25 (D) 1 in 50

10. Generally the life of an earthen road varies in between
(A) 12 to 30 years
(B) 5 to 10 years
(C) 2 to 4 years
(D) Less than one year

11. For proper drainage, the earthen road should have a minimum gradient of—
(A) 1 in 50 (B) 1 in 80
(C) 1 in 100 (D) 1 in 120

12. Earthen roads or low cost are generally constructed as—
(A) National highway (B) State highway
(C) District roads (D) Village roads

13. The cheap and easiest method of soil stabilization is—
(A) Mechanical stabilization
(B) Electrical stabilization
(C) Chemical stabilization
(D) Grouting

14. In a gravel road, the binding material used is—
(A) Cement (B) Lime
(C) Clay (D) Surkhi

15. The small portion of earth left undug in a borrow pit to measure the quantity of earth removed, is known as—
(A) Dead man (B) Embankment
(C) Berm (D) Crest

16. In a gravel road, the amount of clay in the gravel mix varies in between—
    (A) 0 to 10% (B) 15 to 25%
    (C) 30 to 50% (D) 50 to 90%

17. On a WBM road the roller used for rolling purposes is of weight—
    (A) 3 to 5 tonnes (B) 5 to 8 tonnes
    (C) 8 to 10 tonnes (D) 15 to 20 tonnes

18. As light roller is unable to compact a very thick layer and heavy roller crushes the road metal, so thickness of the layer should not exceed—
    (A) 5 cm (B) 10 cm
    (C) 15 cm (D) 30 cm

19. Low cost roads are provided at those places where the intensity of traffic per day does not exceed—
    (A) 200 tonnes (B) 500 tonnes
    (C) 1000 tonnes (D) 2000 tonnes

20. In the final layer, 1 cu metre sand is applied for the road surface for an area of—
    (A) 160 to 180 sq cm
    (B) 160 to 180 sq ft
    (C) 160 to 180 sq yards
    (D) 160 to 180 sq metre

21. Generally while making payment to the contractor for earth work on roads, normal lift should be—
    (A) 1 metre (B) 1·5 metre
    (C) 2·5 metre (D) 32 metre

22. A good subgrade should not have—
    (A) Good strength (B) Good cohesion
    (C) Good capillarity (D) Good drainage

23. Under iron wheeled traffic the best suitable pavement is—
    (A) Flexible pavement
    (B) Rigid pavement
    (C) Earthen roads
    (D) All the above are correct

24. For ordinary earth work on embankment the allowance for settlement of earth is—
    (A) 5% (B) 10%
    (C) 25% (D) 50%

25. As the roots the trees remaining with the ground, decay with passage of time, causing hollow pockets which may result unequal settlement of the road. Hence the grubbing should be done at least upto a depth of—
    (A) 30 cm (B) 50 cm
    (C) 75 cm (D) 90 cm

26. The maximum side slope to be provided in cutting on a road can be—
    (A) 1 : 2 (B) 2 : 1
    (C) 1 : 4 (D) Vertical

27. Sub-base course is provide not to improve—
    (A) Bearing capacity of the subgrade
    (B) Capillary rise of the soil
    (C) Drainage action of the soil
    (D) Durability of the subgrade

28. The thickness of sandy soil to be spread on stone or brick soiling in base course is approximately—
    (A) 50 mm (B) 25 mm
    (C) 75 mm (D) 150 mm

29. In rigid pavements *i.e.*, concrete roads, the course which can be omitted is—
    (A) Subgrade (B) Base course
    (C) Base coat (D) Wearing course

30. The best example of rigid pavement is—
    (A) Concrete road
    (B) Bitumen road
    (C) Gravel road
    (D) Water bound macadam road

31. In tack cost bitumen is used at the rate of—
    (A) 1 kg/m^2 (B) 0·5 kg/m^2
    (C) 5 kg/m^2 (D) 10 kg/m^2

32. The thickness of surface dressing layer in a bituminous road varies in between—
    (A) 0·5 cm to 1 cm (B) 1 cm to 2 cm
    (C) 2 cm to 3 cm (D) 5 cm to 10 cm

33. The Bitumen used in surface dressing for spraying purposes is generally heated to an approximate temperature of—
    (A) 100°C (B) 120°C
    (C) 250°C (D) 180°C

34. As per I. R. C. in surface dressing, the quantity of bitumen used for every 10 m^2 of the surface in first coat on a W.B.M. road is—
    (A) 17 to 19·5 kg (B) 10 to 12 kg
    (C) 12 to 17 kg (D) 19 to 22 kg

35. Generally the premix carpet laid in India is of thickness—
(A) 15 cm (B) 10 cm
(C) 5 to 7·5 cm (D) 2 cm

36. Generally the thickness of asphaltic concrete adopted is—
(A) 2 cm (B) 3 cm
(C) 4 cm (D) 5 cm

37. As per I.R.C. specification in the construction of premix carpet, the grade of bitumen to be used should be of penetration—
(A) 40/50 (B) 80/100
(C) 180/200 (D) 16/120

38. A per I.R.C. the difference in temperature in between aggregate and binder should not be—
(A) More than 14°C (B) Less than 14°C
(C) Less than 20°C (D) More than 20°C

39. Generally a cement concrete road is disliked because—
(A) Its initial cost is high
(B) It causes glare due to reflected sunlight
(C) It requires long time for curing
(D) It becomes noisy under iron wheeled traffic

40. A cement concrete road is preferred because—
(A) It provides an impervious surface
(B) It provides good visibility during night
(C) It has good strength, durability and long life
(D) It provides good riding surface under all conditions

41. The thickness of cement of concrete slab generally adopted varies in between—
(A) 1 cm to 5 cm (B) 5 cm to 10 cm
(C) 10 cm to 15 cm (D) 15 cm to 20 cm

42. If a cement concrete road is properly constructed and maintained, then its expected life will be—
(A) One year (B) 10 years
(C) 20 years (D) 30 to 40 years

43. The width of way, on a concrete road is generally takes as—
(A) > 4·5 m (B) < 4·5 m
(C) 3 to 4 m (D) 2 to 3 m

44. In concrete roads, continuous construction method is preferred as compared to alternate bay method, because—
(A) Number of joints will be less
(B) During construction, half lane can be utilized for the traffic
(C) During rainy season, water may be collect inside the incomplete alternate bay's
(D) In alternate bay method equipment have to be moved back to complete the remaining bay's

45. Compaction of concrete slab is done with the help of—
(A) Light roller (B) Heavy roller
(C) Vibrator (D) Mixer

46. In a concrete road, when subgrade is very poor and the road is subjected to heavy traffic, then—
(A) Sub base is provided
(B) Reinforcement is provided
(C) Soil is stabilized
(D) Asphaltic concrete is provided on the surface

47. To transfer wheel load at expansion joints from one slab to the other—
(A) Tie bars are used
(B) Dowel bars are used
(C) Hanger bars are used
(D) Bitumen filler is used

48. The maximum space between longitudinal joints should be—
(A) 4·5 m (B) 10 m
(C) 15 m (D) 30 m

49. Reinforcement in cement concrete pavement is kept—
(A) 5 cm below from the top level
(B) 5 cm high from the bottom level
(C) In the centre of the slab
(D) In the bottom of the slab

50. Generally coefficient of friction between concrete pavement are subgrade is takes as—
(A) 0·3 to 0·5 (B) 0·5 to 0·8
(C) 0·8 to 1·5 (D) 1·5 to 2

51. The use of water for raising the crops is known as—
(A) Agriculture (B) Irrigation
(C) Water supply (D) Agriculture water

52. Today out of total water consumed for human needs in the world about .... is used for raising the crops.
(A) 50% (B) 60%
(C) 80% (D) 95%

53. At present it is estimated that only about .... of the agriculturable land of the world is receiving assured irrigation facilities.
(A) 15% (B) 30%
(C) 50% (D) 80%

54. The present sources of water are enough to provide irrigation to about .... of the irrigable area of the world.
(A) 15% (B) 30%
(C) 50% (D) 80%

55. Over irrigation may cause excessive rise in the ground water table and thus may be the cause of waterloggings and salt efflorescence of the area—
(A) True (B) False

56. The upper layer of soil fit for agriculture purposes, contains—
(A) Air
(B) Free carbonate and minerals
(C) Moisture
(D) All the above are correct

57. The most important physical properties of the soil which markedly affect growth of the plant are—
(A) Soil texture
(B) Soil structure
(C) Both of the above are correct
(D) None of these are correct

58. If the total dissolved solid in water are up to .... all the waters are considered fit for irrigation purposes.
(A) Zero ppm (B) 100 ppm
(C) 400 ppm (D) 1200 ppm

59. If the total dissolved solid in water exceeds .... then the water is considered unsuitable for irrigation.
(A) 1200 ppm (B) 100 ppm
(C) 400 ppm (D) 2400 ppm

60. If plants are unable to unable to extract sufficient water to meet their growth requirements, then the water content present in soil is termed as—
(A) Permanent wilting point
(B) Moisture equivalent
(C) Wilting coefficient
(D) Both (A) and (B) are correct

61. The area irrigated by unit quantity of water flowing for full growth period of some specified crop is known as the .... for that particular crop.
(A) Base period (B) Delta
(C) Crop period (D) Duty of water

62. The time for which a crop occupies a field to attain its full maturity, is known as—
(A) Duty (B) Base period
(C) Delta (D) Crop period

63. The total depth of water, required to be supplied during the entire period of the crop is known as—
(A) Duty (B) Base period
(C) Delta (D) Crop period

64. The time in days from the instant of showing the crop to that of harvesting is known as—
(A) Duty (B) Base period
(C) Delta (D) Crop period

65. The ratio in between number of days a canal has actually run to the number of days of irrigation period is known as—
(A) Capacity factor
(B) Outlet factor
(C) Full supply coefficient
(D) Time factor

66. It has been observed that only .... of the water supplied from the diversion head work is by the crops.
(A) 70 to 75% (B) 50 to 60%
(C) 28 to 29% (D) 10 to 20%

67. The approximate percentage in loss of water in the field due to surface evaporation, deep percolation and irregular distribution is—
(A) 50% (B) 22%
(C) 15% (D) 28%

68. If proper utilization of technology is made then it is very easily possible to use 100% of water supplied for the growth of crops—
(A) True (B) False

69. Out of the three methods, sub soil method of irrigation is the best one and by this method efficiency can be achieved even—
(A) 100%
(B) 50%
(C) Higher than 80%
(D) Higher than 100%

70. In furrow method of surface irrigation, the depth of furrow is generally made up—
(A) 40 to 50 cm (B) 20 to 30 cm
(C) 10 to 20 cm (D) 5 to 10 cm

71. The innermost part or core having soft tissues found near about the centre of log of a timber tree is called—
(A) Pith (B) Heart wood
(C) Sap wood (D) Bast

72. The layer of liquid material deposited below the bark and outside the sapwood in the log of a timber tree is called—
(A) Pith (B) Heart wood
(C) Sapwood (D) Cambium layer

73. The outer part of log of a timber tree inside the bark which contains living cells is called—
(A) Pith (B) Medullary rays
(C) Sap wood (D) Cambium layer

74. Annual rings represent .... of the tree.
(A) Life (B) Strength
(C) Density (D) Type

75. A piece of timber under 5 cm thick and less than 10 cm in width is called—
(A) Strip (B) Batten
(C) Plank (D) Cant

76. A piece of sawn timber, the dimensions of which does not exceed 5 cm either in breadth or in thickness is called—
(A) Strip (B) Batten
(C) Plank (D) Cant

77. A short log 1·25 m or less in length is called—
(A) Pale (B) Bale
(C) Bolt (D) Post

78. A thin plank generally under 5 cm thick and 10 cm or more in width is called—
(A) Batten (B) Plank
(C) Board (D) Scantling

79. In a well seasoned piece of wood the percentage of moisture present should be—
(A) 0% (B) 5 to 10%
(C) 10 to 12% (D) 30 to 40%

80. When timber is seasoned artificially, *i.e.*, by kiln seasoning then .... is injected under pressure in the chamber for seasoning purpose.
(A) Hot air (B) Stream
(C) Cold air (D) Hot water

81. The radial ruptures in annual rings or splits radiating from the heart and extending towards the sapwood are called—
(A) Radial shake (B) Heart shake
(C) Ring gall (D) Burl

82. The curved unvelling resulted from the growth of sapwood layer on wound left by a falling branch of cut off branch in an irregular manner is called—
(A) Rind gall (B) Burl
(C) Cup shake (D) Upsets

83. While making identification of timber by punched card key system the total number of diaganostic features classified are—
(A) 21 (B) 27
(C) 63 (D) 100

84. The usual thickness of plywood board for 3 ply is—
(A) 1 to 3 mm (B) 1 to 3 cm
(C) 3 to 5 cm (D) 3 to 5 mm

85. The average weight of sandal is—
(A) 915 km/m^3 (B) 785 kg/m^3
(C) 865 kg/m^3 (D) 575 kg/m^3

86. The best wood to be used for superior quality of furniture is—
(A) Kail wood (B) Sandal wood
(C) Shisham wood (D) Teak wood

87. Due to attack of dry rot-timber—
(A) Shrinks
(B) Cracks
(C) Reduces to powder
(D) Swells

88. The fluid paste prepared by dissolving a base into a vehicle along with a colouring pigment is known as—
(A) Paint (B) Varnish
(C) French polish (D) Lacquer

89. Generally a surface is painted in order to—
(A) Protect the surface from weathering effects
(B) Prevent decay of wooden members by wet rot
(C) Provide colourful pleasing and decorative appearances
(D) All the above are correct

90. In an oil paint, the base common in use is—
(A) Red lead (B) White lead
(C) Oxide of iron (D) Whiting

91. The function of a base in an oil paint are—
(A) It makes the film of the paint hard durable and more resistant to abrasion
(B) It gets oxidised in drying and forms a thin tough adhesive durable and water proofing film
(C) It provides glass to the painted surface
(D) All the above are correct

92. Generally in a paint the oil which is used as a vehicle is—
(A) Turpentine oil
(B) Kerosene oil
(C) Linseed oil
(D) All the above are correct

93. The linseed oil dries very quickly and it requires a thinner like turpentine for its proper use, only when it is—
(A) In raw stage
(B) Single boiled
(C) Double boiled
(D) In raw stage and given heat treatment

94. The volatile substances that are added to the paint to make its application easy and smooth is called solvent and generally for it .... is used.
(A) Linseed oil (B) Poppy oil
(C) Nut oil (D) Turpentine oil

95. The paint having white lead or zinc white as base mixed in oil with petroleum spirits and resinous matter is known as—
(A) Enamel paint (B) Emulsion paint
(C) Varnish (D) Lacquer

96. The main constituents of cement paints are—
(A) White cement
(B) Hydrated lime
(C) Hygroscopic salts
(D) All the above are correct

97. The vehicle used in distemper is—
(A) Linseed oil (B) Varnish
(C) Turpentine oil (D) Water

98. Out of the following, the material which dries quickly is—
(A) Paint (B) Polish
(C) Enamel (D) Varnish

99. Aluminium paint protects .... better than any other paint.
(A) Steel (B) Timber
(C) Cement plaster (D) All are correct

100. For coating the buried parts of wood .... is used.
(A) Oil paint (B) Cement paint
(C) Lacquer (D) Bitumen

ANSWERS

1. (B)	2. (A)	3. (D)	4. (A)	5. (D)
6. (A)	7. (C)	8. (B)	9. (D)	10. (C)
11. (D)	12. (D)	13. (A)	14. (C)	15. (A)
16. (B)	17. (C)	18. (C)	19. (B)	20. (D)
21. (B)	22. (C)	23. (A)	24. (B)	25. (D)
26. (D)	27. (B)	28. (B)	29. (C)	30. (A)
31. (B)	32. (C)	33. (D)	34. (A)	35. (D)
36. (C)	37. (B)	38. (A)	39. (A)	40. (C)
41. (D)	42. (D)	43. (A)	44. (B)	45. (C)
46. (B)	47. (B)	48. (A)	49. (A)	50. (D)
51. (B)	52. (C)	53. (A)	54. (C)	55. (A)
56. (D)	57. (C)	58. (C)	59. (A)	60. (D)
61. (D)	62. (B)	63. (C)	64. (D)	65. (D)
66. (C)	67. (B)	68. (B)	69. (B)	70. (B)
71. (A)	72. (D)	73. (C)	74. (A)	75. (A)
76. (B)	77. (C)	78. (C)	79. (C)	80. (B)
81. (B)	82. (A)	83. (C)	84. (D)	85. (A)
86. (D)	87. (C)	88. (A)	89. (D)	90. (B)
91. (A)	92. (C)	93. (C)	94. (D)	95. (A)
96. (D)	97. (D)	98. (B)	99. (A)	100. (D)

Model Set-5

1. While grinding the terrazzo floor, final cutting is done by the carborundum stone of Ne—
(A) 60 (B) 80
(C) 100 (D) 120

2. The boards which are used in wooden flooring are liable to shrinkage and warping and so the maximum width recommended is—
(A) 5 cm (B) 10 cm
(C) 20 cm (D) 30 cm

3. The floor which is best for recommendation in heavy industries, warehouse etc. is—
(A) Granolithic floor (B) Brick floor
(C) Concrete floor (D) Asphaltic floor

4. The thickness of Asphalt floor varies from—
(A) 10 to 15 mm (B) 20 to 30 mm
(C) 25 to 40 mm (D) None of these

5. The wedge shaped bricks or blocks used for the construction of an arch are called as—
(A) Springers (B) Voussoirs
(C) Key stone (D) None of these

6. Springers are those voussoirs of an arch which are provided at—
(A) Its centre (B) Its extremes
(C) Any position (D) All are correct

7. The end supports of an arch are called as—
(A) Piers (B) Abutment
(C) Abutment piers (D) Columns

8. The upper convex surface of an arch is called as—
(A) Intrados (B) Extrados
(C) Soffit (D) Haunch

9. The irregular triangular portion of the arcade, formed between the two arches is called as—
(A) Haunch (B) Spandril
(C) Impost (D) King post

10. The lower half portion of the arch, from the springing to the mid-way to crown from either side is called as—
(A) Haunch (B) Spandril
(C) Extrados (D) Intrados

11. The inclined supports provided in an arch constructed to receive the arch are called as—
(A) Spandril (B) Haunch
(C) Knew back (D) Impost

12. The highest point of an arch is called as
(A) Key stone (B) Crown
(C) Ridge (D) King closer

13. When rise of an arch is more than half of span of the arch and its lower portion is straight instead of circular, then it is called as—
(A) Horse shoe arch
(B) Stilted arch
(C) Lancet arch
(D) Drop arch

14. When rise of an arch is more than the span, then the arch is called as—
(A) Lancet Arch (B) Venetian Arch
(C) Drop Arch (D) Ogee arch

15. The centering for small segmental arches, primarily consists of wooden planks cut to the shape of the arch, are called as—
(A) Arch Ribs (B) Laggings
(C) Struts (D) Ties

16. In any case the bearing of a lintel should not be less than—
(A) 10 cm (B) 15 cm
(C) 20 cm (D) 30 cm

17. In R.C.C. lintels the position of main reinforcement in concrete should be at—
(A) The centre (B) The top
(C) The bottom (D) Any position

18. The covering provided to the building and supported on the top of walls and pillars is known as—
    (A) Tent (B) Roof
    (C) Structural slab (D) Terrace
19. Transverse strength of the wall is......by the roofs provided upon them.
    (A) Increased
    (B) Decreased
    (C) Not affected
    (D) All the above are correct
20. For large spans the most suitable type of covering preferred is—
    (A) Flat roof (B) Slopy roofs
    (C) Shelled roof (D) Domes
21. Flat roof are particularly suitable for tropical places because—
    (A) The top of the flat roof can be used as a terrace
    (B) It equalize the temperature, *i.e.*, rooms under it remains cool in summer and warm in winter
    (C) It increases the architectural appearances of the building
    (D) All the above are correct
22. Pitched roofs are particulary suitable in hilly regions because—
    (A) They are lighter in weight
    (B) It is very easy to repair pitched roof as compared to flat roof
    (C) Of heavy rain fall and snow fall
    (D) All the above are correct
23. The most suitable type of pitched roof for a span of 2·5 metre is—
    (A) Lean to roof
    (B) Couple roof
    (C) Collar roof
    (D) King past truss roof
24. The pitched roof which slopes in all the four directions is named as—
    (A) Shed roof
    (B) Globe roof
    (C) Hzp roof
    (D) Mansard roof
25. The wooden boards fixed along the lower edge of the sloping surface of a pitched roof are named as—
    (A) Common rafter (B) Eave's board
    (C) Barge board (D) Mansard roof
26. The members laid horizontally to support the common rafters and to transmit the load to the trusses walls are called as—
    (A) Purlins (B) Rafters
    (C) Ridge piece (D) Dragon beam
27. The short piece of timber bolted to the rafters of a roof truss fixing the purlins in position are named as—
    (A) Bolt (B) Cleat
    (C) Gable (D) Stopper
28. For a span ranging in between 3·5 m to 4·5 m the most suitable type of roof is—
    (A) Lean to roof (B) Couple roof
    (C) Couple-close roof (D) Collar roof
29. For a span ranging in between 6m to 9m, the most suitable type of pitched roof is—
    (A) King post truss roof
    (B) Queen post truss roof
    (C) Mansard roof truss
    (D) Collar tie roof
30. The inclined braces in a truss to prevent the sagging of the principal rafters are called as—
    (A) King post (B) Tie beam
    (C) Struts (D) Queen post
31. In the king post truss, the member used to prevent the sagging to tie beam and connected with the ridge is called as—
    (A) Collar (B) Rafter
    (C) King post (D) Strut
32. The foot of the queen post is prevented from coming closer by another horizontal beam provided over tie beam and is called as—
    (A) Strairing still (B) Strairing beam
    (C) Strut (D) Queen post
33. In village, the material used for covering the pitched roof is—
    (A) Thatch (B) Country tiles
    (C) A.C. sheets (D) G.I. sheets

34. As a pitched roof covering C.G.I. sheets are used and C.G.I. stands for—
(A) Cement Gauzed iron sheets
(B) Corrugated good iron sheets
(C) Corrugated galvanised sheets
(D) Carbon galvanised iron sheets

35. The standard length of asbestos cement corrugated sheet should be—
(A) 1·52 m
(B) 2·13 m
(C) 3·05 m
(D) All the above are correct

36. In asbestos cement corrugated sheet roofing, the minimum horizontal lap in metres should be—
(A) 0·10 m (B) 2·15 m
(C) 0·20 m (D) 0·30 m

37. To increase the appearance and to hide the unevenness of beams etc. the structure provided below the ceiling is known as—
(A) Terrace (B) Structural slab
(C) False ceiling (D) Ceiling

38. The size of the square used in plywood ceiling should be—
(A) 20 to 30 cm (B) 30 to 50 cm
(C) 50 to 80 cm (D) 100 to 120 cm

39. The thickness of the straw boards used for false ceiling under the name of Sita-Tex boards is—
(A) 10 mm (B) 12 mm
(C) 20 mm (D) 25 mm

40. The covering provided on the walls to improve their appearance, to hide the irregularities of construction .... to provide a base for colour washing of distempering etc. is called as.
(A) Plastering (B) Pointing
(C) White washing (D) Painting

41. Generally white plastering in ordinary buildings, the ratio of cement and sand in cement plaster adopted should be—
(A) 1 : 2 (B) 1 : 3
(C) 1 : 6 (D) 1 : 8

42. Generally while doing time plaster, the ratio in between Lim and surkhi adopted is—
(A) 1 : 2 (B) 1 : 4
(C) 1 : 6 (D) 1 : 8

43. In a truss of equal angle sections are used then they should not be—
(A) Less than 50 × 50 × 6 mm
(B) More than 50 × 50 × 6 mm
(C) Less than 100 × 100 × 10 mm
(D) More than 100 × 100 × 10 mm

44. In a truss, the thickness of gusset plate at any joint should not be—
(A) More than 6 mm
(B) More than 10 mm
(C) Less than 6 mm
(D) Less than 10 mm

45. In a truss, the rivets of nominal diameter .... should not be used.
(A) More than 12 mm
(B) More than 20 mm
(C) Less than 12 mm
(D) Less than 16 mm

46. The total number of varieties of timber which are used for structural purposes are—
(A) 4 (B) 12
(C) 87 (D) 101

47. The fibre saturation point at which the weight of moisture in the wood is in between .... of oven dry weight of wood.
(A) 10 to 15% (B) 15 to 20%
(C) 21 to 28% (D) 30 to 35%

48. When the low durability timber are used on outside locations, the permissible stresses for all grades of timber found are multiplied by—
(A) 1·20 (B) 1·01
(C) 0·90 (D) 0·80

49. The average unit weight of sal is 865 kg/m^3 provided the moisture content in the wood should be—
(A) 20% (B) 12%
(C) 6% (D) Zero%

50. The permissible for grade and timber should be .... when compression is parallel to the grains are timber used is inside the location.
(A) 120 kg/cm^2 (B) 106 kg/cm^2
(C) 70 kg/cm^2 (D) 46 kg/cm^2

51. When the slenderness ratio of a wooden column is ……… then the columns are known as short columns.
(A) Greater than 11 (B) Less than 11
(C) Greater than 5 (D) Less than 5

52. In solid wood columns, the slenderness ratio should not exceed—
(A) 11 (B) 25
(C) 35 (D) 50

53. In a wooden section subjected to tensile forces the area removed by notches should not be—
(A) Less than 25% (B) More than 25%
(C) Less than 33% (D) More than 33%

54. Built up wooden beams are formed by connecting a number of smaller beams together by bolts, screw or spikes. These bolts are staggered and spaced longitudinally at a distance ……… the depth of beams.
(A) Less than four times
(B) Less than eight times
(C) More than four times
(D) More than eight times

55. The minimum width of the beam or any flexural member shall not be less than ……… or 1/50 of the span whichever is greater.
(A) 50 cm (B) 20 cm
(C) 10 cm (D) 5 cm

56. When wooden beams are supported directly on masonry or concrete then a minimum of ……… as bearing length should be provided.
(A) 20 cm (B) 15 cm
(C) 7·5 cm (D) 10 cm

57. In a cantilever beam the maximum deflection of the beam should not be greater than—
(A) $\frac{1}{180}$ (B) $\frac{1}{120}$
(C) $\frac{1}{360}$ (D) $\frac{1}{240}$

58. The wooden beams joined together with steel beams either by means of bolts or screws are known as—
(A) Notched beams
(B) Conjugate beams
(C) Flitched beams
(D) Reinforced beams

59. The distance from end of timber to the centre of bolt is known as end distance. When the load is compressive, the end distance should be ……… the diameter of bolt.
(A) 4 times (B) 5 times
(C) 7 times (D) 2·5 times

60. Nail joints in soft timber are safe and economical to transmit the load up to—
(A) 500 kg (B) 1000 kg
(C) 2500 kg (D) 5000 kg

61. On a hill road, the radius of curve should not be less than—
(A) 100 m (B) 60 m
(C) 30 m (D) 15 cm

62. A curve having concavity on the inner edge of a hill road is known as—
(A) Salient curve
(B) Re-entrant curve
(C) Hair pin curve
(D) Transition curve

63. Out of the three possible sections of a road, the most economical sections of the road is that when it will be—
(A) Totally in cutting
(B) Totally in filling
(C) Partly in cutting and partly in filling
(D) Nothing can be said

64. The drains constructed on up slope of hill sides are known as—
(A) Cross drains (B) Under drains
(C) Catch water drains (D) Side drains

65. When land slide occurs due to the movement within the displaced mass, then it is named as—
(A) Fall
(B) Slide
(C) Flow
(D) Complex land slide

66. Sub-surface drainage of a road means removal of rain water from—
(A) Surface of the road
(B) Sub-grade or sub surface of the road
(C) Surface and sub surface of the road
(D) Soil of permanent land of the road

67. The efficiency and life of a road can be considerably increased by providing—
(A) Good road drainage system
(B) Good road surface
(C) Good soil information bed
(D) Good material in road construction

68. In order to keep sub-surface sufficiently dry, the sub soil water-level should remain below subgrade of the road by at least—
(A) 15 to 30 cm (B) 50 to 80 cm
(C) 100 to 120 cm (D) 150 to 300 cm

69. At places of heavy rainfall side drains to be provided on a road should be—
(A) Triangular in section
(B) Trapezoidal in section
(C) Rectangular in section
(D) Circular in section

70. In order to reduce endanger of floods, the height of a road embankment should be—
(A) 60 cm above the highest flood level of the area
(B) 200 cm above the highest flood of the area
(C) At a level equal to the highest flood level of the area
(D) 100 cm above the ground water level of the area

71. In side drains, the free board should remain at least—
(A) 25 cm below from the road surface
(B) 50 cm high from the road surface
(C) 25 cm high from the sub-soil water level
(D) 50 cm below from the sub-soil water level

72. In case of embankments, the distance between side drains and toe of the embankments should be at least—
(A) 50 cm (B) 100 cm
(C) 200 cm (D) 300 cm

73. In order to keep the road surface dry, side drains are constructed. The drains filled with filter materials are known as—
(A) Jelly drains (B) Sub-soil drains
(C) French drains (D) Scuppers

74. If a road has to be provided on cutting, then side drains should be—
(A) On one side of the road
(B) On both sides of the road
(C) In middle of the road
(D) On both sides and middle of the road

75. The water which penetrates into the ground will continue to flow until it meets with some impermeable strata. The top surface of this flowing water will be called as—
(A) Water table
(B) Afflux
(C) Free water level
(D) Highest flood level

76. The depth of water table below the ground level depends upon—
(A) Geological conditions
(B) Level of the ground
(C) Source of seepage
(D) Quantity of water

77. 10 to 12 cm diameter holes provided in a retaining structure to drain off the seepage flow are known as—
(A) Weep holes (B) Honey combing
(C) Scuppers (D) Saucer drains

78. In order the subgrade material should not get into the filter material, the permeability ratio between filter material and sub-grade material should be—
(A) Less than 5 (B) Greater than 5
(C) Less than 10 (D) Greater than 10

79. The care and planting of trees, *i.e.*, tree culture is known as—
(A) Lopping (B) Pruning
(C) Land scape (D) Arboriculture

80. Gold Mahar, Mango, Jamun and Malina trees are suitable for—
(A) Loamy soil (B) Clayey soil
(C) Sandy soil (D) Usar soil

81. Shisham, Safedsin, Arroo Kanji trees are suitable in—
(A) Loamy soil (B) Clayey soil
(C) Usar soil (D) Sandy soil

82. Neem, Imli, Dhak etc. are most suitable in—
(A) Loamy soil (B) Clayey soil
(C) Usar soil (D) Sandy soil

83. For plantation of trees, the approximate size of the pit to be dugged should be—
(A) 30 × 30 × 30 cm
(B) 60 × 60 × 60 cm
(C) 1 × 1 × 1 m
(D) 2 × 2 × 2 m

84. After plantation, trees should be looked for at least—
(A) 3 to 6 months
(B) 6 to 12 months
(C) 12 to 24 months
(D) 24 to 36 months

85. Normally trees are planted longitudinally at a distance of—
(A) 1 to 2 m (B) 2 to 5 m
(C) 8 to 10 m (D) 10 to 15 m

86. The distance between trunk of a tree and centre of carriage away should be in between—
(A) 3 to 5 m (B) 5 to 8 m
(C) 8 to 10 m (D) 10 to 15 m

87. The removal of dead wood, broken branches and other branches which turn back towards centre of the tree or closely parallel branches is known as—
(A) Felling (B) Lopping
(C) Pruning (D) Hoeing

88. In case of water logged areas, trees should be planted on—
(A) The ground levelled low
(B) Sufficiently raised ground
(C) The ground level
(D) On the ground parallel to the road surface

89. After planting and sapling, the soil is refilled up to a depth of—
(A) 15 cm below ground level
(B) 15 cm above ground level
(C) Equal to the ground level
(D) 30 cm above ground level

90. The lopping shall be carried out either out either at the end of cold weather or at the end of rains.
(A) True (B) False

91. Excessive deformation in foundation course of a flexible pavement is known as—
(A) Base course failure
(B) Wearing course failure
(C) Subgrade failure
(D) Pavement failure

92. The longitudinal depressions or cuts formed on flexible road pavements are known as—
(A) Pot holes (B) Ruts
(C) Patches (D) Map-cracking

93. The circular isolated depressions formed on flexible road pavements are known as—
(A) Ruts (B) Pot holes
(C) Frost heaving (D) Map-cracking

94. Through maintenance of earthen road should be carried out—
(A) Before commencement of rains
(B) After rains
(C) Before summer
(D) Before winter

95. In cement concrete roads, the repair work of cracks should be done just on the end of—
(A) Summer season (B) Rainy season
(C) Winter season (D) Spring season

96. Patching should be done just—
(A) After rainy season
(B) After summer season
(C) Before winter season
(D) Before rainy season

97. The waveness in bituminous road surface is due to—
(A) Mud pumping
(B) Non-uniformity of aggregates
(C) Settlement of the subgrades
(D) Sliding of side slopes

98. The chipping of pavement at joints due to the entry of grits or defective construction is known as—
(A) Scalling
(B) Spalling
(C) Blow ups
(D) Disintegration of pavements

99. Structural of concrete slab is due to—
    (A) Inadequate slab thickness
    (B) Incorrect spacing of joints
    (C) Inadequate compaction
    (D) Inadequate foundation

100. Shear failure cracking in a road pavement is due to—
    (A) Settlement in the filling material
    (B) Sliding of side slopes
    (C) Weakness of the pavements
    (D) Excessive wearing of the pavements

ANSWERS

1. (D)	2. (B)	3. (A)	4. (B)	5. (B)
6. (B)	7. (B)	8. (B)	9. (B)	10. (A)
11. (C)	12. (B)	13. (B)	14. (A)	15. (A)
16. (B)	17. (C)	18. (B)	19. (A)	20. (B)
21. (D)	22. (D)	23. (A)	24. (C)	25. (C)
26. (A)	27. (B)	28. (C)	29. (A)	30. (C)
31. (C)	32. (A)	33. (A)	34. (B)	35. (D)
36. (B)	37. (C)	38. (B)	39. (A)	40. (A)
41. (B)	42. (C)	43. (A)	44. (C)	45. (D)
46. (C)	47. (C)	48. (D)	49. (B)	50. (C)
51. (B)	52. (D)	53. (B)	54. (A)	55. (D)
56. (C)	57. (A)	58. (C)	59. (A)	60. (C)
61. (D)	62. (A)	63. (C)	64. (C)	65. (C)
66. (B)	67. (A)	68. (C)	69. (B)	70. (A)
71. (A)	72. (C)	73. (A)	74. (B)	75. (A)
76. (A)	77. (A)	78. (B)	79. (D)	80. (B)
81. (D)	82. (C)	83. (C)	84. (D)	85. (D)
86. (D)	87. (C)	88. (B)	89. (A)	90. (A)
91. (A)	92. (B)	93. (B)	94. (B)	95. (A)
96. (A)	97. (B)	98. (B)	99. (A)	100. (C)